T0310558

Computational Chemotaxis Models for Neurodegenerative Disease

Computational Chemotaxis Models for Neurodegenerative Disease

William E Schiesser

Lehigh University, USA

 World Scientific

W JERSEY · LONDON · SINGAPORE · BEIJING · SHANGHAI · HONG KONG · TAIPEI · CHENNAI · TOKYO

Published by

World Scientific Publishing Co. Pte. Ltd.

5 Toh Tuck Link, Singapore 596224

USA office: 27 Warren Street, Suite 401-402, Hackensack, NJ 07601

UK office: 57 Shelton Street, Covent Garden, London WC2H 9HE

Library of Congress Cataloging-in-Publication Data

Names: Schiesser, W. E., author.

Title: Computational chemotaxis models for neurodegenerative disease /
   by William E. Schiesser (Lehigh University, USA).

Description: Hackensack, NJ : World Scientific, 2017. |
   Includes bibliographical references and index.

Identifiers: LCCN 2017002352| ISBN 9789813207455 (hardcover : alk. paper) |
   ISBN 9789813208919 (softcover : alk. paper)

Subjects: LCSH: Nervous system--Degeneration. | Alzheimer's disease. |
   Differential equations, Partial. | Chemotaxis.

Classification: LCC RC365 .S35 2017 | DDC 616.8/0471--dc23

LC record available at https://lccn.loc.gov/2017002352

British Library Cataloguing-in-Publication Data

A catalogue record for this book is available from the British Library.

Typeset by Stallion Press
Email: enquiries@stallionpress.com

Printed in Singapore

Contents

Preface

The mathematical model presented in this book, based on partial differential equations (PDEs), is suggested for a quantitative analysis of neurodegenative disease (ND), e.g., Alzheimer's disease (AD). It is a representation of basic phenomena (mechanisms) for diffusive transport and biochemical kinetics that provides the spatiotemporal distribution of components which could explain the evolution of ND, and is offered with the intended purpose of providing a small step toward the understanding, and possible treatment of ND. Additional background is provided in the brief introduction of Chapter 1.

The format and emphasis of the presentation is based on the following elements:

1. A statement of the PDE system, including initial conditions (ICs), boundary conditions (BCs) and the model parameters;
2. Algorithms for the calculation of numerical solutions of the PDE system with a minimum of mathematical formality;
3. A set of R routines for the calculation of numerical solutions, including a detailed explanation of all of the sections of the code. The R routines can be executed after a straightforward download of R, an open-source scientific computing system available from the Internet;
4. Presentation of the numerical solutions, particularly in graphical (plotted) format to enhance the visualization of the solution; and

5. Summary and conclusions concerning the principal results from the model that might serve as the basis for a next step in the modeling of ND.

In other words, a methodology for numerical PDE modeling is presented that is flexible, open-ended and readily implemented on modest computers. If the reader is interested in an alternate model, it might possibly be implemented by: (1) modifying and/or extending the current model (for example, by adding terms to the PDEs or adding additional PDEs), or (2) using the reported routines as a prototype for the model of interest. These suggestions illustrate an important feature of computer-based modeling, that is, the readily available procedure of numerically experimenting with a model

The author would welcome comments about the reported model and how it might be applied, and possibly extended, for an enhanced understanding of ND. It is offered as only a first step toward the resolution of this urgent medical problem.

W. E. Schiesser
Bethlehem, PA USA
01FEB17

Chapter 1

Introduction

This book is predicated on the assumption that a chemotaxis model[1] is a plausible mathematical model for the dynamics of neurodegenerative disease, particularly Alzheimer's disease (AD). The intent of this book is to provide a generic chemotaxis model for neurodegenerative disease, with emphasis on AD, that can be used to develop new models as experimental observations (data) and understanding through continuing research become available. A central feature of the generic model is an implementation as a series of computer routines written in R[2] that are explained (documented) in detail.

The model includes the production (e.g., from invasion) and diffusion of a pathogen, subsequently termed a *cell*,[3] and the consequent production of proteins, subsequently termed *chemicals*, that contribute to the chemotaxis movement of cells [2].

---

[1] *Chemotaxis* is an established term that denotes movement (taxis), e.g., of biochemicals, biological cells, viruses, under the influence of chemical gradients (chemo) [14]. The mathematical modeling of chemotaxis can take various forms to describe the diffusion resulting from gradients of the system components. An extensive literature has developed pertaining to chemotaxis. A brief introduction to the mathematical modeling of diffusion fluxes is given in the chapter appendix.

[2] R is a quality, open source scientific programming system that is available for download through an Internet connection, http://www.r-project. org/, http://cran.fhcrc.org/.

[3] The pathogen might, for example, be a bacterial cell or a virus.

A distinguishing feature of the chemicals is that they are both attractants and repellents for cell movement [7, 8, 9, 10, 13]. The principal feature of the model solution is the spatiotemporal distribution of the cells and chemicals.

The cells can initiate an immune response with local infection and inflammation [6]. The immune response in turn produces chemicals such as β-amyloid proteins (Aβ) [3], Interleuken-6 (IL-6) and TNF-α (tumor necrosis factor-α) that serve as attractants and repellents [9]. A series of feedback effects leads to a spatiotemporal distribution of cells and chemicals, with subsequent mis-folded protein aggregation [1] (e.g., senile plaques) that can cause neuron damage exhibited as neurodegenerative disease such as AD.

These ideas of attractant-repellent chemotaxis are described in detail in the subsequent presentation of the model and the R implementation. Since spatial effects are included, as well as time, the model is stated as a system of partial differential equations (PDEs). The numerical solution of these equations as implemented in the R routines is based on the methods of lines (MOL),[4] a general approach to the numerical integration of PDEs [12].

We now proceed to the statement of the model, followed by its computer implementation.

---

[4]Briefly, MOL is a numerical procedure (algorithm) for PDEs in which the spatial (boundary value) derivatives are replaced with an algebraic approximation such as finite differerences (used in this work), finite elements, finite volumes, spectral, Galerkin, weighted-residual or least squares representations. The resulting ordinary differential equations (ODEs) in an initial value variable, typically time, are then integrated (solved) with a library ODE integrator or a programmed (coded) initial value algorithm. The solution to the approximating ODEs constitute a numerical solution to the original PDE problem system.

References

[1] Chalub, F., et al (2006), Model hierarchies for cell aggregation by chemotaxis, *Mathematical Models and Methods in Applied Science*, **16**, no. 7S, 1173–1197

[2] Edelstein-Keshet, L, and A. Spiros (2002), Exploring the formation of Alzheimer's disease senile plaques in silico, *J. Theoretical Biology*, **216**, 301–326

[3] Gandy, S., and F.L. Heppner (2005), Breaking up (amyloid) is hard to do, *PLoS Medicine*, **2**, no. 12.

[4] Hillen, T., and K.J. Painter (2009), A user's guide to PDE models for chemotaxis, *Journal of Mathematical Biology*, **58**, 183

[5] Keller, E.F., and L.A. Segel (1971), Traveling bands of chemotactic bacteria: A theoretical analysis, *Journal of Theoretical Biology*, **30**, 235–248

[6] Kumar, D.V.K, et al (2016), Amyloid-β peptide protects against microbial infection in mouse and worm models of Alzheimer's disease, *Science Translational Medicine*, **8**, no. 340, 1–15

[7] Li, Y., and Y. Li (2016), Blow-up of nonradial solutions to attraction-repulsion chemotaxis system in two dimensions, *Nonlinear Analysis: Real World Applications*, **30**, 170–183

[8] Lin, K., and C. Mu (2016), Global existence and convergence to steady states for an attraction-repulsion chemotaxis system, *Nonlinear Analysis: Real World Applications*, **31**, 630–643

[9] Luca, M., A. Chavez-Ross, L. Edelstein-Keshet and A. Mogilner (2003), Chemotactic signaling, microglia, and Alzheimer's disease senile plaques: Is there a connection?, *Bulletin of Mathematical Biology*, **65**, 693–730

[10] Luri, I.K., and L. Li (2010), Mathematical modeling for the pathogenesis of Alzheimer's disease, *PLoS One*, **5**, no. 12.

[11] Murray, J.D. (2003), *Mathematical Biology, II: Spatial Models and Biomedical Applications*, Third Edition

[12] Schiesser, W.E. (2014), *Differential Equation Analysis in Biomedical Science and Engineering: Partial Differential Equation Applications in R*, John Wiley and Sons, Hoboken, NJ

[13] Wang, Y. (2016), A quasilinear attraction-repulsion chemotaxis system of parabolic-elliptic type with logistic source, *J. Mathematical Analysis and Applications*, **441**, 259–292

[14] Wang, W. (2017), A diffusive virus infection dynamic model with nonlinear functional response, absorption effect and chemotaxis, *Communications in Nonlinear Science and Numerical Simulation*, **42**, 585–606

Appendix A1.1: Diffusion Models

Several chemotaxis diffusion models with one and two components are briefly reviewed here. The two component models can be classified as chemotaxis in the sense that the diffusion of one of the components is determined in part by the second component.

The one component models are listed in the following table which gives the flux of a single component, q_{x,u_1}, as a function of the component concentration or population, u_1, for a 1D, Cartesian coordinate system with spatial variable x.

One component, u_1

no.	flux	comments
(1)	$q_{x,u_1} = -D\dfrac{\partial u_1}{\partial x}$	Fick's first law
(2)	$q_{x,u_1} = -Du_1\dfrac{\partial u_1}{\partial x}$	nonlinear diffusion
(3)	$q_{x,u_1} = -f(u_1)\dfrac{\partial u_1}{\partial x}$	arbitrary function $f(u_1)$

u_1 is the concentration of the (single) component. Model (1) is the conventional, linear Fickian model wih diffusivity D (a constant). Models $(2), (3)$ are nonlinear extensions of (1) (for example, note the product $u_1 \dfrac{\partial u_1}{\partial x}$ in (2)). $f(u_1)$ is a positive function that can be used to define the diffusivity as a function of the concentration u_1.

The fluxes of $(1), (2), (3)$ can be used in a conservation PDE[5]

$$\frac{\partial u_1}{\partial t} = -\frac{\partial q_{x,u_1}}{\partial x} \tag{A1.1}$$

For example, using (1), we have

$$\frac{\partial u_1}{\partial t} = -\frac{\partial \left(-D \dfrac{\partial u_1}{\partial x} \right)}{\partial x} = D \frac{\partial^2 u_1}{\partial x^2} \tag{A1.2}$$

which is *Fick's second law* or the *diffusion equation* (also known as *Fourier's second law* in heat transfer).

Similar substitutions of $(2), (3)$ in eq. (A1.1) gives nonlinear versions of the diffusion equation. Analytical solutions for these nonlinear diffusion equations may not be available, but in principle, they can be integrated numerically.

The two component models are listed in the following table which gives the flux of component 1, q_{x,u_1}, as a function of the components, u_1, u_2, for a 1D, Cartesian coordinate system with spatial variable x.

---

[5]In words, eq. (A1.1) states

rate of accumulation (or depletion) in a differential volume =
net rate into (or out of) the differential volume by diffusion

In the case of eq. (A1.1), the differential volume is $A dx$ where A is a cross sectional area and dx is a Cartesian differential length. Other geometries are possible as expressed in various coordinate systems, e.g., spherical coordinates which are used in the subsequent discussion of the chemotaxis models (in approximate conformity with a human brain).

Two components, u_1, u_2

no.	flux, u_1	comments
(4) $q_{x,u_1} = -f_1(u_2)\dfrac{\partial u_1}{\partial x} + u_1 f_2(u_2)\dfrac{\partial u_2}{\partial x}$		chemotaxis 1 [5]
(5) $q_{x,u_1} = -D_1\dfrac{\partial u_1}{\partial x} + D_2 u_1\dfrac{\partial u_2}{\partial x}$		chemotaxis 2 [11]
(6) $q_{x,u_1} = -D_1\dfrac{\partial u_1}{\partial x} + D_2\left(\dfrac{u_1}{u_2}\right)\dfrac{\partial u_2}{\partial x}$		chemotaxis 3 [11]

Note in particular the application of the gradient $\dfrac{\partial u_2}{\partial x}$ for $u_2(x,t)$ in defining the flux of $u_1(x,t)$.

Chemotaxis has been discussed in an extensive literature. A survey of PDE chemotaxis models is in [4].

Chapter 2

3-PDE Chemotaxis Model

The model PDEs are first stated in generalized coordinates, which can then be specialized to a particular coordinate system that conforms to the physical system, in the present case, spherical coordinates that conform approximately to a human brain.

(2.1) 3-PDE Model, Orthogonal Coordinates

The three PDEs follow.

$$\frac{\partial m}{\partial t} = \mu\nabla^2 m - \chi_1\nabla \cdot (m\nabla c_1) + \chi_2\nabla \cdot (m\nabla c_2) + f(m, c_1, c_2)$$

$$\text{(2.1a)}$$

$$\frac{\partial c_1}{\partial t} = D_1\nabla^2 c_1 + g_1(m, c_1, c_2) \qquad \text{(2.1b)}$$

$$\frac{\partial c_2}{\partial t} = D_2\nabla^2 c_2 + g_2(m, c_1, c_2) \qquad \text{(2.1c)}$$

Variable, function operator, parameter	Interpretation
m	cell concentration[1]
c_1	attractant concentration[1]

---

[1] The interpretation of cells is open to the analyst, e.g., microglia, invasive bacteria. That is, the model is generic and becomes specific with the definition of the type and number of PDEs, ICs, BCs, parameters. Similarly, attractant and repellent can be interpreted by the analyst.

c_2	repellent concentration[1]
t	time
$\nabla \cdot$	divergence of a vector
∇	gradient of a scalar
$f(m, c_1, c_2)$ $g_1(m, c_1, c_2)$ $g_2(m, c_1, c_2)$	volumetric source functions for m, c_1, c_2, respectively
μ, χ_1, χ_2	functions for diffusion and chemotaxis

Table 2.1: Variables, functions, operators, parameters in eqs. (2.1)

$\mu, \chi_1, \chi_2, D_1, D_2$ are functions (initially, constant parameters) to be specified. and $f(m, c_1, c_2)$, $g_1(m, c_1, c_2)$, $g_2(m, c_1, c_2)$ are volumetric source functions to be specified. The physical interpretation of the various LHS and RHS terms in eqs. (2.1) is discussed subsequently.

Eqs. (2.1) are now specialized to spherical coordinates, (r, θ, ϕ).

(2.2) Vector Operators in Spherical Coordinates

From [1, 2], the spatial differential operators (denoted with ∇) in spherical coordinates are given in Tables 2.1, 2.2.

$$
\left[
\begin{array}{c}
[\nabla]_r = \dfrac{1}{r^2} \dfrac{\partial}{\partial r}(r^2) \\[3mm]
[\nabla]_\theta = \dfrac{1}{r \sin \theta} \dfrac{\partial}{\partial \theta}(\sin \theta) \\[3mm]
[\nabla]_\phi = \dfrac{1}{r \sin \theta} \dfrac{\partial}{\partial \phi}
\end{array}
\right]
$$

Table 2.1: $\nabla \cdot$ (divergence of a vector, spherical coordinates)

$$\begin{bmatrix} [\nabla]_r = \dfrac{\partial}{\partial r} \\[2em] [\nabla]_\theta = \dfrac{1}{r}\dfrac{\partial}{\partial \theta} \\[2em] [\nabla]_\phi = \dfrac{1}{r \sin \theta}\dfrac{\partial}{\partial \phi} \end{bmatrix}$$

Table 2.2: ∇ (gradient of a scalar, spherical coordinates)

The source of inflammation that leads to ND, e.g., AD,[2] is located in the neighborhood of $r = 0$. Symmetry with respect to the angles (θ, ϕ), is assumed so that only derivatives with respect to r appear in the PDEs.

(2.3) 1D, 3-PDE Model, Spherical Coordinates

With angular variations neglected, eqs. (2.1) become (with the vector differential operators $\nabla\cdot, \nabla$ from Tables 2.1, 2.2)

$$\frac{\partial m}{\partial t} = \mu \frac{1}{r^2}\frac{\partial}{\partial r}\left(r^2 \frac{\partial m}{\partial r}\right)$$

$$-\chi_1 \frac{1}{r^2}\frac{\partial}{\partial r}\left(r^2 m \frac{\partial c_1}{\partial r}\right) + \chi_2 \frac{1}{r^2}\frac{\partial}{\partial r}\left(r^2 m \frac{\partial c_2}{\partial r}\right) + f(m, c_1, c_2)$$

$$(2.2a)$$

$$\frac{\partial c_1}{\partial t} = D_1 \frac{1}{r^2}\frac{\partial}{\partial r}\left(r^2 \frac{\partial c_1}{\partial r}\right) + g_1(m, c_1, c_2) \qquad (2.2b)$$

$$\frac{\partial c_2}{\partial t} = D_2 \frac{1}{r^2}\frac{\partial}{\partial r}\left(r^2 \frac{\partial c_2}{\partial r}\right) + g_2(m, c_1, c_2) \qquad (2.2c)$$

---

[2]ND $\leftrightarrow$ neurodegenerarive disease, AD $\leftrightarrow$ Alzheimer's disease.

Expanding the radial groups

$$\frac{\partial m}{\partial t} = \mu \left(\frac{\partial^2 m}{\partial r^2} + \frac{2}{r} \frac{\partial m}{\partial r} \right)$$

$$-\chi_1 \left(m \frac{\partial^2 c_1}{\partial r^2} + \frac{\partial m}{\partial r} \frac{\partial c_1}{\partial r} + \frac{2}{r} m \frac{\partial c_1}{\partial r} \right)$$

$$+\chi_2 \left(m \frac{\partial^2 c_2}{\partial r^2} + \frac{\partial m}{\partial r} \frac{\partial c_2}{\partial r} + \frac{2}{r} m \frac{\partial c_2}{\partial r} \right) + f(m, c_1, c_2) \qquad (2.3a)$$

$$\frac{\partial c_1}{\partial t} = D_1 \left(\frac{\partial^2 c_1}{\partial r^2} + \frac{2}{r} \frac{\partial c_1}{\partial r} \right) + g_1(m, c_1, c_2) \qquad (2.3b)$$

$$\frac{\partial c_2}{\partial t} = D_2 \left(\frac{\partial^2 c_2}{\partial r^2} + \frac{2}{r} \frac{\partial c_2}{\partial r} \right) + g_2(m, c_1, c_2) \qquad (2.3c)$$

For $r = 0$, eqs. (2.3) become[3]

$$\frac{\partial m}{\partial t} = 3\mu \frac{\partial^2 m}{\partial r^2}$$

$$-\chi_1 \left(3m \frac{\partial^2 c_1}{\partial r^2} + \frac{\partial m}{\partial r} \frac{\partial c_1}{\partial r} \right)$$

$$+\chi_2 \left(3m \frac{\partial^2 c_2}{\partial r^2} + \frac{\partial m}{\partial r} \frac{\partial c_2}{\partial r} \right) + f(m, c_1, c_2) \qquad (2.4a)$$

$$\frac{\partial c_1}{\partial t} = 3D_1 \frac{\partial^2 c_1}{\partial r^2} + g_1(m, c_1, c_2) \qquad (2.4b)$$

$$\frac{\partial c_2}{\partial t} = 3D_2 \frac{\partial^2 c_2}{\partial r^2} + g_2(m, c_1, c_2) \qquad (2.4c)$$

---

[3] At $r = 0$, the first-order radial groups are indeterminate. Application of l'Hospital's rule ([5], p298) gives $\frac{2}{r} \frac{\partial m}{\partial r} = 2 \frac{\partial^2 m}{\partial r^2}$, $\frac{2}{r} \frac{\partial c_1}{\partial r} = 2 \frac{\partial^2 c_1}{\partial r^2}$, $\frac{2}{r} \frac{\partial c_2}{\partial r} = 2 \frac{\partial^2 c_2}{\partial r^2}$.

Eqs. (2.3) and (2.4) are second order in r and they therefore each require two boundary conditions (BCs).

$$\frac{\partial m(r = 0, t)}{\partial r} = 0 \qquad (2.5a)$$

$$\mu \frac{\partial m(r = r_0, t)}{\partial r}$$

$$-\chi_1 m(r = r_0, t) \frac{\partial c_1(r = r_0, t)}{\partial r}$$

$$+\chi_2 m(r = r_0, t) \frac{\partial c_2(r = r_0, t)}{\partial r}$$

$$= 0 \qquad (2.5b)$$

$$\frac{\partial c_1(r = 0, t)}{\partial r} = \frac{\partial c_1(r = r_0, t)}{\partial r} = 0 \qquad (2.5c,d)$$

$$\frac{\partial c_2(r = 0, t)}{\partial r} = \frac{\partial c_2(r = r_0, t)}{\partial r} = 0 \qquad (2.5e,f)$$

Eqs. (2.5) are termed *homogeneous, Neumann* BCs. They are Neumann since they specify the derivative of the solutions at the boundaries.[4] They are homogeneous since the derivatives are zero. Physically, eqs. (2.5) specify no diffusion or flux of the cells and chemicals through the boundaries ($r = r_0$). They are therefore termed *no-flux* or *impermeable* BCs. In the case of eqs. (2.5a), (2.5c) and (2.5e), the zero derivatives result from *symmetry* of the solutions.

Eqs. (2.3) and (2.4) are first order in t, and they therefore each require one initial condition (IC).

$$m(r, t = 0) = h_m(r); \ c_1(r, t = 0) = h_1(r); \ c_2(r, t = 0) = h_2(r)$$
$$(2.6a,b,c)$$

h_m, h_1, h_2 are functions to be specified.

---

[4]BCs that specify the PDE dependent variables at the boundaries are termed *Dirichlet*.

For the first example to follow, the volumetric source terms in eqs. (2.3) and (2.4) are taken as $f(m, c_1, c_2) = r_1 e^{-r_2 r^2}$ (a Gaussian function with specified constants r_1, r_2), $g_1(m, c_1, c_2) = a_1 m - b_1 c_1$, $g_2(m, c_1, c_2) = a_2 m - b_2 c_2$. The functions g_1, g_2 represent first order reaction terms based on m (a forward or production reaction) and c_1, c_2 (for reverse or consumption reactions) with specified constants a_1, a_2, b_1, b_2. These functions are specific selections as suggested in [3], but they can be of essentially any form in the routines to follow.

We conclude this section with a brief explanation (interpretation) of the terms in eqs. (2.2), which are component (mass) balances on a differential shell of thickness dr and volume $4\pi r^2 dr$.[5]

For eq. (2.2a),

- $\mu \dfrac{1}{r^2} \dfrac{\partial}{\partial r} \left(r^2 \dfrac{\partial m}{\partial r} \right)$: Conventional Fick's second law for the net diffusion into or out the differential volume $4\pi r^2 dr$.

- $-\chi_1 \dfrac{1}{r^2} \dfrac{\partial}{\partial r} \left(r^2 m \dfrac{\partial c_1}{\partial r} \right)$: Chemotaxis diffusion into or out of the differential volume. The diffusion flux is given by

$$\chi_1 m \frac{\partial c_1}{\partial r}$$

which is a departure from the conventional Fick's first law with the following features:

1. The rate of diffusion of the cells is proportional to the gradient of a chemical, $\dfrac{\partial c_1}{\partial r}$, rather than the gradient of the cells, $\dfrac{\partial m}{\partial r}$.

2. The sign of the flux is opposite to that of Fick's first law, so the cells move in the direction of *increasing*

---

[5] A detailed derivation of mass balances in Cartesian, cylindrical and spherical coordinates is given in [4], Appendix 1.

chemical concentration, c_1. Thus, the chemical is termed an *attractant* and the term models *chemotaxis* (movement in response to a chemical gradient).

3. The rate of diffusion is also proportional to the concentration of cells, m. The product $m\dfrac{\partial c_1}{\partial r}$ is therefore nonlinear (since it depends on the two dependent variables m, c_1).

4. The parameter χ_1 serves as a type of diffusivity.

- $+\chi_2\dfrac{1}{r^2}\dfrac{\partial}{\partial r}\left(r^2 m\dfrac{\partial c_2}{\partial r}\right)$: Chemotaxis diffusion into or out of the differential volume in response to a second chemical, c_2. The diffusion flux is given by

$$-\chi_2 m\frac{\partial c_2}{\partial r}$$

which has the features

1. The rate of diffusion of the cells is proportional to the gradient of a second chemical, $\dfrac{\partial c_2}{\partial r}$, rather than the gradient of the cells, $\dfrac{\partial m}{\partial r}$.

2. The sign of the flux corresponds to Fick's first law, so the cells move in the direction of *decreasing* chemical concentration, c_2. Thus, the chemical is a *repellent* and this term acts in the opposite way to the first term for chemotaxis.

3. The rate of diffusion is also proportional to the concentration of cells, m. The product $m\dfrac{\partial c_2}{\partial r}$ is therefore nonlinear (since it depends on the two dependent variables m, c_2).

4. The parameter χ_2 serves as a type of diffusivity.

- The net chemotaxis depends on the additive effect of the two terms (with χ_1 and χ_2).
- $+f(m, c_1, c_2)$: A volumetric source terms for the cells.

- $\dfrac{\partial m}{\partial t}$: The accumulation (when positive) or depletion (when negative) of cells in the differential volume. The variation with t depends on the additive effect of the conventional and chemotaxis diffusion of the cells, and the volumetric source term, as expressed by the RHS of eq. (2.2a).

Eq. (2.2a) is a nonlinear diffusion equation for which a numerical MOL algorithm is programmed in the routines discussed next.

For eq. (2.2b),

- $D_1 \dfrac{1}{r^2} \dfrac{\partial}{\partial r}\left(r^2 \dfrac{\partial c_1}{\partial r} \right)$: Conventional (Fick's second law) diffusion of the first chemical with concentration c_1 (the diffusion is in the direction of decreasing c_1). The diffusivity is D_1.
- $+g_1(m, c_1, c_2)$: A volumetric source terms for the first chemical.
- $\dfrac{\partial c_1}{\partial t}$: The accumulation (when positive) or depletion (when negative) of the first chemical in the differential volume. The variation with t depends on the additive effect of the conventional diffusion and the source term as expressed by the RHS of eq. (2.2b).

The interpretation of eq. (2.2c) is the same as for eq. (2.2b) with c_1 replaced by c_2.

Eqs. (2.2) are a system of nonlinear PDEs for which an analytical solution would be difficult to derive, particularly when variable properties $(\mu, \chi_1, \chi_2, D_1, D_2)$ and various forms of the source terms are considered. The numerical approach detailed in the following discussion is straightforward.

We next consider numerical values for the parameters in eqs. (2.3) to (2.6). This then completes the specification of the chemotaxis model so that a numerical solution can be programmed and computed as discussed in the next chapter.

(2.4) Model Parameters

The model is based on *cgs* (centimeter, grams, seconds) units. The parameters in eqs. (2.3) to (2.5) are listed in Table 2.3 (with $1\,M = 1\,mol/liter = 1\,mol/1000\,cm^3$).

Parameter	Value	Units
μ	3×10^{-9}	cm^2/sec
χ_1	1	$cm^2/M\text{-sec}$
χ_2	1	$cm^2/M\text{-sec}$
D_1	10^{-7}	cm^2/sec
D_2	10^{-7}	cm^2/sec
a_1	1.0×10^{-12}	$M\text{-}cm^3/cells\text{-sec}$
a_2	1.2×10^{-12}	$M\text{-}cm^3/cells\text{-sec}$
b_1	5×10^{-4}	$1/sec$
b_2	5×10^{-4}	$1/sec$
r_0	10	cm
r_1	10^4	$cells/cm^3\text{-sec}$
r_2	1	$1/cm^2$

Table 2.3: Numerical parameters for eqs. (2.3) to (2.6)

The numerical values in Table 2.3 were selected to test the model, and are considered to be approximately order-of-magnitude estimates. If the model (eqs. (2.3) to (2.6)) is used and extended, more precise values may be available from the analysis of observed solution properties and data.

In the next chapter, the method of lines (MOL) coding of eqs. (2.2) to (2.6) is presented through a series of transportable R routines.

References

[1] Bird, R.B., W.E. Stewart and E.N. Lightfoot (2002), *Transport Phenomena*, 2nd ed, John Wiley and Sons, Hoboken, NJ, p836

[2] Soetaert, K., J. Cash, and F. Mazzia (2012), *Solving Differential Equations in R*, Springer-Verlag, Heidelberg, Germany, p140–141

[3] Luca, M., A. Chavez-Ross, L. Edelstein-Keshet and A. Mogilner (2003), Chemotactic signaling, microglia, and Alzheimer's disease senile plaques: Is there a connection?, *Bulletin of Mathematical Biology*, **65**, 693-730

[4] Schiesser, W.E. (2013), *Partial Differential Equation Analysis in Biomedical Engineering*, Cambridge University Press, Cambridge, UK

[5] Schiesser, W.E. (2014), *Differential Equation Analysis in Biomedical Science and Engineering: Partial Differential Equation Applications in R*, John Wiley and Sons, Hoboken, NJ

Chapter 3

R Routines

In this chapter, we consider the R routines for the numerical method of lines (MOL) solution of eqs. (2.3) to (2.6). Basically, in the MOL, the spatial (boundary value) derivatives in the PDEs are replaced with algebraic approximations, and in the present development, with finite difference (FDs). Then only one independent variable remains, an initial value variable, typically time t, so that a set of ODEs results that approximates the PDEs. This system of ODEs is then integrated numerically by a library initial value ODE integrator.

Since the model MOL/ODEs are central to this approach, we start with the MOL/ODE routine, then consider the main program that calls it.

(3.1) ODE/MOL Routine

The ODE routine, pde_1a, is listed next.

```
  pde_1a=function(t,u,parms){
#
# Function pde_1a computes the t derivative
# vectors of m(r,t),c1(r,t),c2(r,t)
#
# One vector to three vectors
  m=rep(0,nr);c1=rep(0,nr);c2=rep(0,nr);
```

```
  for(i in 1:nr){
    m[i]=u[i];
    c1[i]=u[i+nr];
    c2[i]=u[i+2*nr];
  }
#
# mr,c1r,c2r
  mr=dss006(0,r0,nr, m);
  c1r=dss006(0,r0,nr,c1);
  c2r=dss006(0,r0,nr,c2);
#
# BCs
  mr[1]=0;  mr[nr]=0;
  c1r[1]=0; c1r[nr]=0;
  c2r[1]=0; c2r[nr]=0;
#
# mrr,c1rr,c2rr
  mrr=dss006(0,r0,nr, mr);
  c1rr=dss006(0,r0,nr,c1r);
  c2rr=dss006(0,r0,nr,c2r);
#
# PDEs
  mt=rep(0,nr);c1t=rep(0,nr);c2t=rep(0,nr);
  for(i in 1:nr){
    if(i==1){
      mt[i]=3*mu*mrr[i]-chi1*3*m[i]*c1rr[i]+
                        chi2*3*m[i]*c2rr[i]+
                        f1(r[i]);
      c1t[i]=3*d1*c1rr[i]+a1*m[i]-b1*c1[i];
      c2t[i]=3*d2*c2rr[i]+a2*m[i]-b2*c2[i];
    }
    if(i>1){
      mt[i]=mu*(mrr[i]+(2/r[i])*mr[i])-
        chi1*(m[i]*c1rr[i]+mr[i]*c1r[i]+
```

```
          (2/r[i])*m[i]*c1r[i])+
          chi2*(m[i]*c2rr[i]+mr[i]*c2r[i]+
          (2/r[i])*m[i]*c2r[i])+f1(r[i]);
       c1t[i]=d1*(c1rr[i]+(2/r[i])*c1r[i])+
          a1*m[i]-b1*c1[i];
       c2t[i]=d2*(c2rr[i]+(2/r[i])*c2r[i])+
          a2*m[i]-b2*c2[i];
    }
  }
#
# Three vectors to one vector
  ut=rep(0,3*nr);
  for(i in 1:nr){
    ut[i]      =mt[i];
    ut[i+nr]   =c1t[i];
    ut[i+2*nr]=c2t[i];
  }
#
# Increment calls to pde_1a
  ncall <<- ncall+1;
#
# Return derivative vector
  return(list(c(ut)));
  }
```

Listing 3.1: Routine pde_1a for the MOL solution of eqs. (2.3) to (2.5)

We can note the following details about Listing 3.1.

- The function is defined.

```
   pde_1a=function(t,u,parms){
#
# Function pde_1a computes the t derivative
# vectors of m(r,t),c1(r,t),c2(r,t)
```

The input arguments are:

— t, the current value of t.
— u, the vector of ODE dependent variables, with nr = 101 ODEs (nr is set in the main program discussed subsequently) approximating each of eqs. (2.3) and (2.4), so that u is of length $3(101) = 303$.
— parms, parameters passed to pde_1a, which is unused but must still be declared (a requirement of the ODE integrator discussed subsequently).

The output of pde_1a is the 303-vector of ODE derivatives in t as computed in pde_1a (discussed next).

- u is placed in three vectors, m,c1,c2, to facilitate the programming of eqs. (2.3) and (2.4) in terms of equation-based variables. The vectors are first declared (preallocated) with the rep utility.

```
#
# One vector to three vectors
  m=rep(0,nr);c1=rep(0,nr);c2=rep(0,nr);
  for(i in 1:nr){
    m[i]=u[i];
    c1[i]=u[i+nr];
    c2[i]=u[i+2*nr];
  }
```

- The derivatives $\dfrac{\partial m}{\partial r}, \dfrac{\partial c_1}{\partial r}, \dfrac{\partial c_2}{\partial r}$ are computed by the spatial differentiator dss006.

```
#
# mr,c1r,c2r
  mr=dss006(0,r0,nr, m);
  c1r=dss006(0,r0,nr,c1);
  c2r=dss006(0,r0,nr,c2);
```

dss006 implements sixth-order FDs which give good spatial accuracy for nr = 101 as discussed subsequently.

- BCs (2.5) are set.

```
#
# BCs
   mr[1]=0;   mr[nr]=0;
   c1r[1]=0;  c1r[nr]=0;
   c2r[1]=0;  c2r[nr]=0;
```

Subscripts 1, nr correspond to $r = 0, r_0$. BC (2.5b) reduces to $\dfrac{\partial m(r = r_0, t)}{\partial r} = 0$ from BCs (2.5d), (2.5f).

- The second derivatives $\dfrac{\partial^2 m}{\partial r^2}, \dfrac{\partial^2 c_1}{\partial r^2}, \dfrac{\partial^2 c_2}{\partial r^2}$ are computed by differentiating the first derivatives (*stagewise differentiation*).

```
#
# mrr,c1rr,c2rr
   mrr=dss006(0,r0,nr, mr);
   c1rr=dss006(0,r0,nr,c1r);
   c2rr=dss006(0,r0,nr,c2r);
```

- Eqs. (2.3) and (2.4) are programmed in a **for** that steps through the grid points in r. The derivatives in t (the LHSs of eqs. (2.3) and (2.4)) are placed in three vectors, mt, c1t, c2t, which are first declared with the **rep** utility. For eqs. (2.4), the programming is (note the 3 that distinguishes the $r = 0$ form of the PDEs)

```
#
# PDEs
   mt=rep(0,nr);c1t=rep(0,nr);c2t=rep(0,nr);
   for(i in 1:nr){
     if(i==1){
       mt[i]=3*mu*mrr[i]-chi1*3*m[i]*c1rr[i]+
                        chi2*3*m[i]*c2rr[i]+
```

```
            f1(r[i]);
   c1t[i]=3*d1*c1rr[i]+a1*m[i]-b1*c1[i];
   c2t[i]=3*d2*c2rr[i]+a2*m[i]-b2*c2[i];
 }
```

The programming reflects the particular choice of the volumetric source terms $f(m, c_1, c_2) = r_1 e^{-r_2 r^2}$ (a Gaussian function with specified constants r_1, r_2), $g_1(m, c_1, c_2) = a_1 m - b_1 c_1$, $g_2(m, c_1, c_2) = a_2 m - b_2 c_2$ [2]. The Gaussian function is programmed in function f1(r[i]) discussed subsequently. The constants $\mu, \chi_1, \chi_2, D_1, D_2, a_1, a_2, b_1, b_2$ and nr are defined numerically in the main program discussed subsequently, and are available to pde_1a without any special designation (a feature of R).

- The programming for eqs. (2.3) $(r \neq 0)$ is

```
  if(i>1){
    mt[i]=mu*(mrr[i]+(2/r[i])*mr[i])-
      chi1*(m[i]*c1rr[i]+mr[i]*c1r[i]+
      (2/r[i])*m[i]*c1r[i])+
      chi2*(m[i]*c2rr[i]+mr[i]*c2r[i]+
      (2/r[i])*m[i]*c2r[i])+f1(r[i]);
    c1t[i]=d1*(c1rr[i]+(2/r[i])*c1r[i])+
      a1*m[i]-b1*c1[i];
    c2t[i]=d2*(c2rr[i]+(2/r[i])*c2r[i])+
      a2*m[i]-b2*c2[i];
  }
 }
```

The second } concludes the for in r. The similarity of the PDEs and the programming is a principal feature of the MOL.
- The three derivative vectors mt,c1t,c2t are placed in a single vector ut that is returned to the ODE integrator (called from the main program considered next).

```
#
# Three vectors to one vector
  ut=rep(0,3*nr);
  for(i in 1:nr){
    ut[i]      =mt[i];
    ut[i+nr]   =c1t[i];
    ut[i+2*nr]=c2t[i];
  }
```

- The number of calls to pde_1a is incremented and returned to the main program by the <<- operator.

```
#
# Increment calls to pde_1a
  ncall <<- ncall+1;
```

- The derivative vector ut is returned to the ODE integrator with c (the R vector operator), list (the ODE integrator requires a list) and return.

```
#
# Return derivative vector
  return(list(c(ut)));
```

The } concludes pde_1a.

This concludes the programming of eqs. (2.3), (2.4) and (2.5). The main program that calls pde_1a is considered next.

(3.2) Main Program

The main program is in Listing 3.2.

```
#
#  Three PDE chemotaxis model
#
# Access ODE integrator
```

```
  library("deSolve");
#
# Access functions for numerical solution
  setwd("f:/neuro/chap3");
  source("pde_1a.R");source("f1.R");
  source("dss006.R");
  source("dss008.R");
#
# Select case
  ncase=2;
#
# Parameters
  mu=3.0e-09;d1=1.0e-07;d2=1.0e-07;
  chi1=1;chi2=1;
  a1=1.0e-12;a2=1.2e-12;
  b1=5.0e-04;b2=5.0e-04;
  if(ncase==1){r1=0       ;r2=1;}
  if(ncase==2){r1=1.0e+04;r2=1;}
#
# Grid (in r)
  nr=101;r0=10;
  r=seq(from=0,to=r0,by=(r0-0)/(nr-1));
#
# Independent variable for ODE integration
  t0=0;tf=1.0e+03;nout=6;
  tout=seq(from=t0,to=tf,by=(tf-t0)/(nout-1));
#
# Initial condition (t=0)
  u0=rep(0,3*nr);
  for(i in 1:nr){
    u0[i]      =0;
    u0[i+nr]   =0;
    u0[i+2*nr]=0;
  }
```

```
  ncall=0;
#
# ODE integration
  out=lsodes(y=u0,times=tout,func=pde_1a,
      sparsetype ="sparseint",rtol=1e-6,
      atol=1e-6,maxord=5);
  nrow(out)
  ncol(out)
#
# Arrays for plotting numerical solution
   m=matrix(0,nrow=nr,ncol=nout);
  c1=matrix(0,nrow=nr,ncol=nout);
  c2=matrix(0,nrow=nr,ncol=nout);
  for(it in 1:nout){
    tout[it]=tout[it]/60;
    for(i in 1:nr){
       m[i,it]=out[it,i+1];
      c1[i,it]=out[it,i+1+nr];
      c2[i,it]=out[it,i+1+2*nr];
    }
  }
#
# Display numerical solution
  for(it in 1:nout){
    if((it==1)|(it==nout)){
      cat(sprintf("\n      t      r       m(r,t)
                  c1(r,t)      c2(r,t)\n"));
      for(i in 1:nr){
        cat(sprintf("%6.2f%6.1f%12.3e%12.3e%12.3e\n",
          tout[it],r[i],m[i,it],c1[i,it],c2[i,it]));
      }
    }
  }
#
```

```
# Calls to ODE routine
  cat(sprintf("\n\n ncall = %5d\n\n",ncall));
#
# Plot PDE solutions
#
# m
  par(mfrow=c(1,1));
  matplot(x=r,y=m,type="l",xlab="r (cm)",
          ylab="m(r,t) (cells/cc)",
          xlim=c(0,r0),lty=1,main="m(r,t)",
          lwd=2,col="black");
#
# c1
  par(mfrow=c(1,1));
  matplot(x=r,y=c1,type="l",xlab="r (cm)",
          ylab="c1(r,t) (M = mols/liter)",
          xlim=c(0,r0),lty=1,main="c1(r,t)",
          lwd=2,col="black");
#
# c2
  par(mfrow=c(1,1));
  matplot(x=r,y=c2,type="l",xlab="r (cm)",
          ylab="c2(r,t) (M = mols/liter)",
          xlim=c(0,r0),lty=1,main="c2(r,t)",
          lwd=2,col="black");
```

Listing 3.2: Main program for the MOL solution of eqs. (2.3)
to (2.5)

We can note the following details about Listing 3.2.

- The library of ODE integrators, deSolve, is accessed[1]

---

[1] deSolve is a library of some 20 quality ODE integrators, both stiff and nonstiff. For the integration of the ODEs programmed in pde_1a of Listing 3.1, lsodes is called in the main program of Listing 3.2.

```
#
#  Three PDE chemotaxis model
#
# Access ODE integrator
  library("deSolve");
#
# Access functions for numerical solution
  setwd("f:/neuro/chap3");
  source("pde_1a.R");source("f1.R");
  source("dss006.R");
  source("dss008.R");
```

The routines required for the MOL solution of eqs. (2.3) and (2.4), and called in pde_1a, are accessed with the source utility. The setwd (set working directory) will require editing for the local computer (note the use of / rather than the usual \). f1 is the function for the cell volumetric source term f_1 in eqs. (2.3a) and (2.4a) (discussed next). dss008 is an alternative to dss006 used in pde_1a that can be used to study the effect of the finite difference (FD) approximations of the derivatives in r in eqs. (2.3) and (2.4).

• The parameters (constants) in eqs. (2.3) and (2.4) are defined numerically (and passed to pde_1a of Listing 3.1).

```
#
# Select case
  ncase=2;
#
# Parameters
  mu=3.0e-09;d1=1.0e-07;d2=1.0e-07;
  chi1=1;chi2=1;
  a1=1.0e-12;a2=1.2e-12;
  b1=5.0e-04;b2=5.0e-04;
  if(ncase==1){r1=0        ;r2=1;}
  if(ncase==2){r1=1.0e+04;r2=1;}
```

Two cases are programmed. For `ncase=1`, $r_1 = 0$ in $f(m, c_1, c_2) = r_1 e^{-r_2 r^2}$ so that this source term is zero. Consequently, with zero ICs (eqs. (3.6)), there is nothing to drive the solution away from zero, which should be observed in the output. This apparently trivial case is worth executing since if the solution moves away from zero, a programming error exists that requires correction (this check is indicated in the output that follows).

For `ncase=2`, $r_1 = 10^4$ so that the Gaussian function is not zero, and the solution responds to it (physically, cells are produced in the neighborhood of $r = 0$, possibly from an inflammation). The output is examined in detail in the subsequent discussion.

As stated in Chapter 2, the numerical values of the parameters in Table 2.3 and programmed here were selected to test the model, and are considered to be approximately order-of-magnitude estimates. These constants can then be altered to experiment with the model.[2]

- The spatial grid is defined, consisting of 101 points over the interval $r = 0 \leq r \leq r = r_0 = 10$ using the `seq` operator. Thus, $r = 0, 0.1, \ldots, 10$ cm which is approximately the spatial scale for a human brain.[3]

```
#
# Grid (in r)
```

---

[2]As a word of caution, numerical values for the parameters might be selected that cause the routines to stop executing, possibly due to a failure of the numerical ODE integrator. In other words, a guarantee in advance of the successful execution of the routines is not possible. This type of parameter sensitivity is to be expected, particularly for nonlinear models such as eqs. (2.3) and (2.4).

[3]With a spatial scale of 10 cm, the cell and chemical concentrations are spatially averaged, and the resolution of individual senile plaques, for example, is not attempted (this would require a spatial scale of 10 μm, i.e., microns [2]).

```
nr=101;r0=10;
r=seq(from=0,to=r0,by=(r0-0)/(nr-1));
```

- The interval in t is defined as $0 \le t \le 10^3$ with 6 output points. Thus, the solution is displayed at $t = 0, 200, \ldots, 1000$ sec.

```
#
# Independent variable for ODE integration
t0=0;tf=1.0e+03;nout=6;
tout=seq(from=t0,to=tf,by=(tf-t0)/(nout-1));
```

- The ICs (2.6) are homogeneous (zero) so that the solution is the response to the Gaussian volumetric source term in eqs. (2.3a) and (2.4a). The total number of ICs corresponds to the number of ODEs (303).

```
#
# Initial condition (t=0)
u0=rep(0,3*nr);
for(i in 1:nr){
   u0[i]       =0;
   u0[i+nr]   =0;
   u0[i+2*nr]=0;
}
ncall=0;
```

The counter for the calls to pde_1a is also initialized.
- The 303 ODEs (defined in pde_1a of Listing 3.1) are integrated by lsodes.

```
#
# ODE integration
out=lsodes(y=u0,times=tout,func=pde_1a,
    sparsetype ="sparseint",rtol=1e-6,
    atol=1e-6,maxord=5);
nrow(out)
ncol(out)
```

The input arguments are:

— u0: Vector of ICs. The length of u0 informs `lsodes` of the number of ODEs to be integrated (303).

— `tout`: Vector of 6 output values of t (including $t = 0$).

— `pde_1a`: ODE routine of Listing 3.1.

— `sparsetype ="sparseint"`: Specification of sparse matrix integration, i.e., the $303 \times 303 = 91809$ element ODE Jacobian matrix is defined as a sparse matrix. This large number (91809) demonstrates the utility of sparse matrix processing since most of the elements are zero.

— `rtol=1e-6,atol=1e-6`: Relative and absolute error tolerances of the ODE integration. These are the default values for `lsodes`, but they are programmed explicitly to demonstrate these values. The use of the error tolerances in the numerical integration to adjust the integration step (h refinement) is discussed subsequently.

— `maxord=5`: Maximum order of the integration algorithm. This order changes automatically, starting with order one (at $t = 0$) for the implicit Euler method (p refinement), in an attempt to meet the error tolerances. The algorithm remains operational for stiff ODE systems (rather than switching between stiff and nonstiff methods as in `lsoda`, the default integrator).

`y,times,func,sparsetype,rtol,atol,maxord` are reserved names used by `lsodes`.

The numerical ODE solution is returned in array `out`, which has dimensions `out(6,303+1)` for the `nout=6` values of t and the 303 ODEs. The offset +1 occurs since the values of t in `tout` are also returned in `out`.

The dimensions of `out` are displayed by `nrow,ncol` to confirm the expected dimensions (6×304).

• The numerical solution is placed in three arrays, `m,c1,c2`, for subsequent numerical and graphical (plotted) display. The offset of +1 in i+1,i+1+nr,i+1+2*nr is required since the values of t are also returned in out by lsodes. The utility matrix is used to declare the arrays. t is converted from sec to min by division of 60.

```
#
# Arrays for plotting numerical solution
  m=matrix(0,nrow=nr,ncol=nout);
  c1=matrix(0,nrow=nr,ncol=nout);
  c2=matrix(0,nrow=nr,ncol=nout);
  for(it in 1:nout){
    tout[it]=tout[it]/60;
    for(i in 1:nr){
      m[i,it]=out[it,i+1];
      c1[i,it]=out[it,i+1+nr];
      c2[i,it]=out[it,i+1+2*nr];
    }
  }
```

- The numerical solution of eqs. (2.3) and (2.4) is displayed at $t = 0$ (to check the ICs) and $t = 1000$ (at the end of the solution) to conserve space.

```
#
# Display numerical solution
  for(it in 1:nout){
    if((it==1)|(it==nout)){
      cat(sprintf("\n     t      r       m(r,t)
                c1(r,t)     c2(r,t)\n"));
      for(i in 1:nr){
        cat(sprintf("%6.2f%6.1f%12.3e%12.3e%12.3e\n",
            tout[it],r[i],m[i,it],c1[i,it],c2[i,it]));
      }
    }
  }
```

Two `fors` are used to step through t and r.

- The number of calls to **pde_1a** is displayed as a measure of the computational effort required to compute the solution.

```
#
# Calls to ODE routine
  cat(sprintf("\n\n ncall = %5d\n\n",ncall));
```

- $m(r, t), c_1(r, t), c_2(r, t)$ (the solutions to eqs. (2.3) to (2.6)) are plotted against r with t as a parameter by the `matplot` utility. The plots are each defined as a 1×1 matrix of plots with `par(mfrow=c(1,1))`.

```
#
# Plot PDE solutions
#
# m
  par(mfrow=c(1,1));
  matplot(x=r,y=m,type="l",xlab="r (cm)",
          ylab="m(r,t) (cells/cc)",
          xlim=c(0,r0),lty=1,main="m(r,t)",
          lwd=2,col="black");
#
# c1
  par(mfrow=c(1,1));
  matplot(x=r,y=c1,type="l",xlab="r (cm)",
          ylab="c1(r,t) (M = mols/liter)",
          xlim=c(0,r0),lty=1,main="c1(r,t)",
          lwd=2,col="black");
#
# c2
  par(mfrow=c(1,1));
  matplot(x=r,y=c2,type="l",xlab="r (cm)",
          ylab="c2(r,t) (M = mols/liter)",
          xlim=c(0,r0),lty=1,main="c2(r,t)",
          lwd=2,col="black");
```

The arguments of `matplot` are essentially self explanatory when considering the graphical output (e.g., Figs. 4.1a, 4.2a, 4.3a). Additional documentation is available by entering `help(matplot)` at the R prompt.

This completes the discussion of the main program of Listing 3.2. The function `f1` for the Gaussian volumetric source term in eqs. (2.3a), (2.4a) follows.

(3.3) Subordinate Routine

```
  f1=function(r){
#
# Function f1 computes the inhomogeneous
# volumetric source term of the m(r,t) PDE
#
  f1=r1*exp(-r2*r^2);
#
# Return f1
  return(c(f1));
  }
```

Listing 3.3: Function `f1` for the volumetric source term in eqs. (2.3a), (2.4a)

Function `f1` is a straightforward implementation of the function $f(m, c_1, c_2) = r_1 e^{-r_2 r^2}$. The constants `r1,r2` are defined numerically in the main program of Listing 3.2 and passed to `f1` without any special designation.

For `ncase=1`, the function is zero (`r1=0`) which leads to $\frac{\partial m}{\partial t} = \frac{\partial c_1}{\partial t} = \frac{\partial c_2}{\partial t} = 0$, i.e., the solutions remain at the homogeneous ICs. This case is considered first as a test of the coding of Listings 3.1, 3.2 and 3.3 (a departure from the zero ICs would indicate a programming error).

Finally, the two library differentiation routines, `dss006,` `dss008,` are included in `source` terms in the main program.

```
source("dss006.R");
source("dss008.R");
```

These routines are not discussed here to conserve space, but are documented in [4]. The choice of **dss006** or **dss008** demonstrates any differences in the solutions from variations in the order of the spatial finite differencing (a type of p refinement with $p = 6$ for **dss006** and $p = 8$ for **dss008**).

This completes the programming of equations (2.3) to (2.6). The output from the code is considered in the next chapter.

References

[1] Li, Y., and Y. Li (2016), Blow-up of nonradial solutions to attraction-repulsion chemotaxis system in two dimensions, *Nonlinear Analysis: Real World Applications*, **30**, 170–183

[2] Luca, M., A. Chavez-Ross, L. Edelstein-Keshet and A. Mogilner (2003), Chemotactic signaling, microglia, and Alzheimer's disease senile plaques: Is there a connection?, *Bulletin of Mathematical Biology*, **65**, 693–730

[3] Luri, I.K., and L. Li (2010), Mathematical modeling for the pathogenesis of Alzheimer's disease, *PLoS One*, **5**, no. 12.

[4] Schiesser, W.E. (2014), *Differential Equation Analysis in Biomedical Science and Engineering: Partial Differential Equation Applications in R*, John Wiley and Sons, Hoboken, NJ

Chapter 4

Model Output

The model output from the code in Listings 3.1, 3.2 and 3.3 is presented and discussed in this chapter.

(4.1) PDE Solution Vector

The first model output is for `ncase=1` in Listing 3.2 for which $r_1 = 0$ so that the volumetric source term $f(m, c_1, c_2) = r_1 e^{-r_2 r^2} = 0$ and all of the LHS derivatives in t of eqs. (2.3) and (2.4) are zero. The three PDE dependent variables, $m(r, t)$, $c_(r, t)$, $c_2(r, t)$, are displayed in numerical and graphical format.

```
[1] 6
```

```
[1] 304
```

t	r	m(r,t)	c1(r,t)	c2(r,t)
0.00	0.0	0.000e+00	0.000e+00	0.000e+00
0.00	0.1	0.000e+00	0.000e+00	0.000e+00
0.00	0.2	0.000e+00	0.000e+00	0.000e+00
0.00	0.3	0.000e+00	0.000e+00	0.000e+00
0.00	0.4	0.000e+00	0.000e+00	0.000e+00
0.00	0.5	0.000e+00	0.000e+00	0.000e+00
.			.	
.			.	
.			.	

35

Output for r = 0.6 to 9.4 removed

.   .

.   .

.   .

0.00	9.5	0.000e+00	0.000e+00	0.000e+00
0.00	9.6	0.000e+00	0.000e+00	0.000e+00
0.00	9.7	0.000e+00	0.000e+00	0.000e+00
0.00	9.8	0.000e+00	0.000e+00	0.000e+00
0.00	9.9	0.000e+00	0.000e+00	0.000e+00
0.00	10.0	0.000e+00	0.000e+00	0.000e+00

t	r	m(r,t)	c1(r,t)	c2(r,t)
16.67	0.0	0.000e+00	0.000e+00	0.000e+00
16.67	0.1	0.000e+00	0.000e+00	0.000e+00
16.67	0.2	0.000e+00	0.000e+00	0.000e+00
16.67	0.3	0.000e+00	0.000e+00	0.000e+00
16.67	0.4	0.000e+00	0.000e+00	0.000e+00
16.67	0.5	0.000e+00	0.000e+00	0.000e+00

.   .

.   .

.   .

Output for r = 0.6 to 9.4 removed

.   .

.   .

.   .

16.67	9.5	0.000e+00	0.000e+00	0.000e+00
16.67	9.6	0.000e+00	0.000e+00	0.000e+00
16.67	9.7	0.000e+00	0.000e+00	0.000e+00
16.67	9.8	0.000e+00	0.000e+00	0.000e+00
16.67	9.9	0.000e+00	0.000e+00	0.000e+00
16.67	10.0	0.000e+00	0.000e+00	0.000e+00

ncall =   352

Table 4.1a: Abbreviated numerical output for `ncase=1`

We can note the following details for this output.

- The dimensions of the output matrix out are $6 \times 303 + 1 = 304$ in accordance with the discussion of Listing 3.2.
- The ICs are zero (at t=0). Although this check may seem obvious, it is worthwhile since if the ICs are not correct, the solution will not be correct (the ICs are the starting point for the solution).
- The solutions, $m(r, t), c_1(r, t), c_2(r, t)$, from eqs. (2.3) and (2.4) remain at zero to $t = 1000$ sec $= 16.67$ min. The reader can confirm that this should happen by considering the RHS terms of eqs. (2.3) and (2.4), and keeping in mind that all of the derivatives in r are zero. Also, homogeneous BCs (2.5) are consistent with the ICs so that a discontinuity between the ICs and BCs does not occur which could a change in the solutions.
- The computational effort is modest as expected since the solutions do not change, ncall = 352.

The invariance of the solutions is confirmed in Figs. 4.1a, 4.1b and 4.1c (produced from the calls to matplot in Listing 3.2).

   If now ncase=1 is changed to ncase=2 in the main program of Listing 3.2, the resulting solution (abbreviated) is in Table 4.1b.

```
[1] 6

[1] 304
```

t	r	m(r,t)	c1(r,t)	c2(r,t)
0.00	0.0	0.000e+00	0.000e+00	0.000e+00
0.00	0.1	0.000e+00	0.000e+00	0.000e+00
0.00	0.2	0.000e+00	0.000e+00	0.000e+00
0.00	0.3	0.000e+00	0.000e+00	0.000e+00
0.00	0.4	0.000e+00	0.000e+00	0.000e+00
0.00	0.5	0.000e+00	0.000e+00	0.000e+00

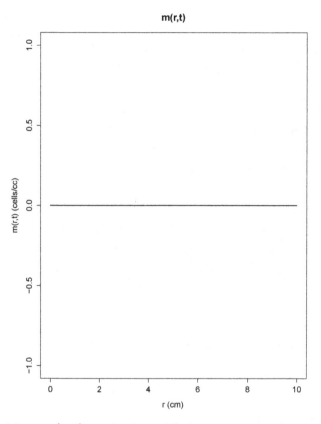

Figure 4.1a: $m(r, t)$ against r with t as a parameter, `ncase=1`

.   .

.   .

.   .

Output for r = 0.6 to 9.4 removed

.   .

.   .

.   .

0.00	9.5	0.000e+00	0.000e+00	0.000e+00
0.00	9.6	0.000e+00	0.000e+00	0.000e+00
0.00	9.7	0.000e+00	0.000e+00	0.000e+00
0.00	9.8	0.000e+00	0.000e+00	0.000e+00

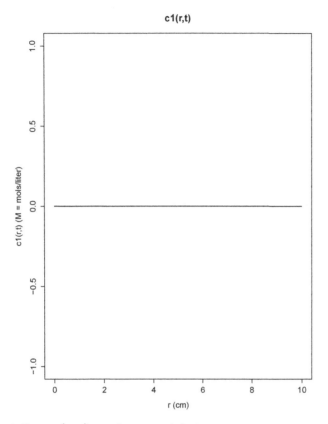

Figure 4.1b: $c_1(r,t)$ against r with t as a parameter, `ncase=1`

0.00	9.9	0.000e+00	0.000e+00	0.000e+00
0.00	10.0	0.000e+00	0.000e+00	0.000e+00

t	r	m(r,t)	c1(r,t)	c2(r,t)
16.67	0.0	5.077e+06	3.023e-03	3.628e-03
16.67	0.1	5.061e+06	3.006e-03	3.607e-03
16.67	0.2	5.013e+06	2.954e-03	3.545e-03
16.67	0.3	4.934e+06	2.870e-03	3.444e-03
16.67	0.4	4.823e+06	2.754e-03	3.305e-03
16.67	0.5	4.682e+06	2.609e-03	3.131e-03

.   .

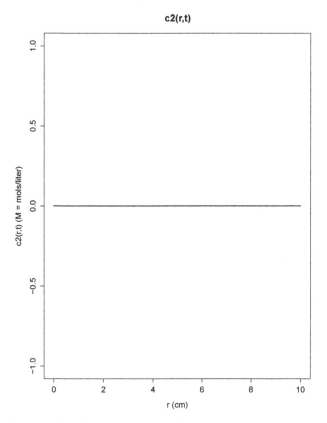

Figure 4.1c: $c_2(r, t)$ against r with t as a parameter, `ncase=1`

```
            .                          .

            .                          .

        Output for r = 0.6 to 9.4 removed

            .                          .

            .                          .

            .                          .

16.67    9.5    6.389e-33    2.791e-42    3.349e-42
16.67    9.6    9.462e-34    4.140e-43    4.968e-43
16.67    9.7    1.374e-34    6.055e-44    7.265e-44
16.67    9.8    1.945e-35    7.733e-45    9.280e-45
16.67    9.9    2.729e-36    1.235e-45    1.481e-45
```

```
16.67   10.0    1.155e-37   -2.206e-45   -2.648e-45

ncall =    444
```

Table 4.1b: Abbreviated numerical output for **ncase=2**

We can note the following details for this output.

- The dimensions of **out** are again 6×304 (since the only change is in r_1).
- The ICs are zero (at $t = 0$).
- The solution, $m(r, t), c_1(r, t), c_2(r, t)$, from $t = 0$ to $t = 1000 \sec = 16.67 \min$ varies in response to the volumetric source term $f(m, c_1, c_2) = r_1 e^{-r_2 r^2}$ in eqs. (2.3a), (2.4a) (with $r_1 = 10^4$).
- The computational effort is modest, **ncall = 444**, even though the solution now changes substantially with r and t. In other words, **lsodes** is computationally efficient for this 303-ODE system.

The complete solutions can be visualized in Figs. 4.2a, 4.2b, 4.2c.
    We can note the following details about Figs. 4.2a, 4.2b, 4.2c.

- The ICs are $m(r, t = 0) = c_1(r, t = 0) = c_2(r, t = 0) = 0$.
- The cell density, $m(r, t)$ and chemical concentrations, $c_1(r, t)$, $c_2(r, t)$, increase with t. In fact, they would continue to increase indefinitely since cells continue to be added with the source term $f(m, c_1, c_2) = r_1 e^{-r_2 r^2}$. A variant of the model might therefore be to place a limit in time of this source term, e.g.,

$$f(m, c_1, c_2) = r_1 e^{-r_2 r^2} > 0, \quad 0 \le t \le t_s$$
$$f(m, c_1, c_2) = r_1 e^{-r_2 r^2} = 0, \quad t > t_s$$

where t_s is a prescribed stopping time. This switching can easily be accomplished in **pde_1a** of Listing 3.1 since the running value of t is available as the first argument of **pde_1a**. More

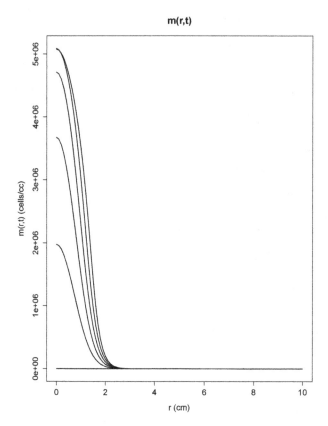

Figure 4.2a: $m(r, t)$ against r with t as a parameter, `ncase=2`

gradual reductions of the source term with t (rather than a discontinuous change to zero) could also be programmed in `pde_1a` as a function of t. The same can be said for an increase in the source term, that is, it could be considered to build up slowly with t (by programming the increase in the source term as a function of time). The time variation of the Gaussian source term is considered subsequently (in Chapter 5).

• The solutions appear to satisfy BCs (2.5) at $r = 0, r = r_0 = 10$.

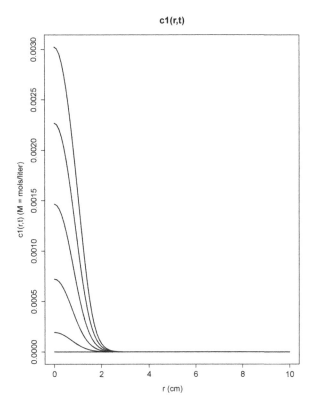

Figure 4.2b: $c_1(r, t)$ against r with t as a parameter, `ncase=2`

- The solutions are smooth (in r) so that 101 grid points in r appear to be adequate. The accuracy of the solutions is discussed in more detail later.
- The cell density $m(r, t)$ and chemical concentrations $c_1(r, t), c_2(r, t)$ remain in the neighborhood of $r = 0$ with increasing t. In particular, if the cells and chemicals are expected to move (disperse, diffuse) toward the boundary $r = r_0$, this feature can be developed by Fickian diffusion, that is, by increasing the diffusivities μ, D_1, D_2 in eqs. (2.3) and (2.4). This aspect of the modeling is left for the reader to study.

The model response in Figs. 4.2a, 4.2b and 4.2c is relatively complex and the question then arises as to how this solution

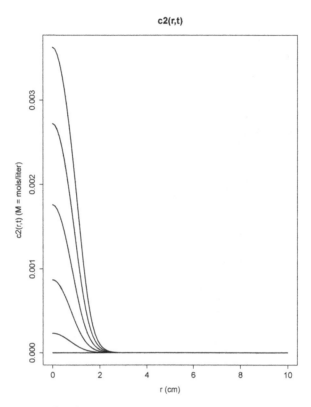

Figure 4.2c: $c_2(r, t)$ against r with t as a parameter, `ncase=2`

results from eqs. (2.3) to (2.6). This question can be studied by examining the various terms in eqs. (2.3) to (2.6). This has the important feature of assessing the relative contributions of the various terms. This more detailed study of the model (rather than just considering the PDE solutions) is developed in the following sections.

(4.2) PDE Solution Vector with t Derivative Vectors

The main program of Listing 3.2 is extended so that the t derivative vectors of eqs. (2.3) and (2.4), i.e., the MOL approximations of $\dfrac{\partial m}{\partial t}$, $\dfrac{\partial c_1}{\partial t}$, $\dfrac{\partial c_2}{\partial t}$, are computed and displayed numerically and graphically.

```
#
#  Three PDE chemotaxis model
#
# Access ODE integrator
  library("deSolve");
#
# Access functions for numerical solution
  setwd("f:/neuro/chap4");
  source("pde_1b.R");source("f1.R");
  source("dss006.R");
  source("dss008.R");
#
# Select case
  ncase=2;
#
# Parameters
  mu=3.0e-09;d1=1.0e-07;d2=1.0e-07;
  chi1=1;chi2=1;
  a1=1.0e-12;a2=1.2e-12;
  b1=5.0e-04;b2=5.0e-04;
  if(ncase==1){r1=0        ;r2=1;}
  if(ncase==2){r1=1.0e+04;r2=1;}
#
# Grid (in r)
  nr=101;r0=10;
  r=seq(from=0,to=r0,by=(r0-0)/(nr-1));
#
# Independent variable for ODE integration
  t0=0;tf=1.0e+03;nout=6;
  tout=seq(from=t0,to=tf,by=(tf-t0)/(nout-1));
```

```
#
# Initial condition
  u0=rep(0,3*nr);
  for(i in 1:nr){
    u0[i]      =0;
    u0[i+nr]   =0;
    u0[i+2*nr]=0;
  }
  ncall=0;
#
# ODE integration
  out=lsodes(y=u0,times=tout,func=pde_1b,
      sparsetype ="sparseint",rtol=1e-6,
      atol=1e-6,maxord=5);
  nrow(out)
  ncol(out)
#
# Arrays for plotting numerical solution
   m=matrix(0,nrow=nr,ncol=nout);
  c1=matrix(0,nrow=nr,ncol=nout);
  c2=matrix(0,nrow=nr,ncol=nout);

  for(it in 1:nout){
    tout[it]=tout[it]/60;
    for(ir in 1:nr){
       m[ir,it]=out[it,ir+1];
      c1[ir,it]=out[it,ir+1+nr];
      c2[ir,it]=out[it,ir+1+2*nr];
    }
  }
#
# Display numerical solution
  for(it in 1:nout){
    if((it==1)|(it==nout)){
```

```
    cat(sprintf("\n        t      r      m(r,t)
                c1(r,t)      c2(r,t)\n"));
    for(ir in 1:nr){
      cat(sprintf("%6.2f%6.1f%12.3e%12.3e%12.3e\n",
      tout[it],r[ir],m[ir,it],c1[ir,it],c2[ir,it]));
    }
  }
}
#
# Calls to ODE routine
  cat(sprintf("\n\n ncall = %5d\n\n",ncall));
#
# Plot PDE solutions
#
# m
  par(mfrow=c(1,1));
  matplot(x=r,y=m,type="l",xlab="r (cm)",
          ylab="m(r,t) (cells/cc)",
          xlim=c(0,r0),lty=1,main="m(r,t)",
          lwd=2,col="black");
#
# c1
  par(mfrow=c(1,1));
  matplot(x=r,y=c1,type="l",xlab="r (cm)",
          ylab="c1(r,t) (M = mols/liter)",
          xlim=c(0,r0),lty=1,main="c1(r,t)",
          lwd=2,col="black");
#
# c2
  par(mfrow=c(1,1));
  matplot(x=r,y=c2,type="l",xlab="r (cm)",
          ylab="c2(r,t) (M = mols/liter)",
          xlim=c(0,r0),lty=1,main="c2(r,t)",
          lwd=2,col="black");
```

```
#
# LHS terms in PDEs
    mr=matrix(0,nrow=nr,ncol=nout);
   c1r=matrix(0,nrow=nr,ncol=nout);
   c2r=matrix(0,nrow=nr,ncol=nout);
   mrr=matrix(0,nrow=nr,ncol=nout);
  c1rr=matrix(0,nrow=nr,ncol=nout);
  c2rr=matrix(0,nrow=nr,ncol=nout);
    mt=matrix(0,nrow=nr,ncol=nout);
   c1t=matrix(0,nrow=nr,ncol=nout);
   c2t=matrix(0,nrow=nr,ncol=nout);
#
# mr,c1r,c2r
  for(it in 1:nout){
  mr[,it]=dss006(0,r0,nr, m[,it]);
  c1r[,it]=dss006(0,r0,nr,c1[,it]);
  c2r[,it]=dss006(0,r0,nr,c2[,it]);
#
# BCs
   mr[1,it]=0;   mr[nr,it]=0;
  c1r[1,it]=0; c1r[nr,it]=0;
  c2r[1,it]=0; c2r[nr,it]=0;
#
# mrr,c1rr,c2rr
   mrr[,it]=dss006(0,r0,nr, mr[,it]);
  c1rr[,it]=dss006(0,r0,nr,c1r[,it]);
  c2rr[,it]=dss006(0,r0,nr,c2r[,it]);
#
# PDEs
  for(i in 1:nr){
    if(i==1){
      mt[i,it]=3*mu*mrr[i,it]-chi1*3*m[i,it]*
               c1rr[i,it]+chi2*3*m[i,it]*
               c2rr[i,it]+f1(r[i]);
```

```
    c1t[i,it]=3*d1*c1rr[i,it]+a1*m[i,it]-
              b1*c1[i,it];
    c2t[i,it]=3*d2*c2rr[i,it]+a2*m[i,it]-
              b2*c2[i,it];
  }
  if(i>1){
    mt[i,it]=mu*(mrr[i,it]+(2/r[i])*mr[i,it])-
            chi1*(m[i,it]*c1rr[i,it]+mr[i,it]*
            c1r[i,it]+(2/r[i])*m[i,it]*c1r[i,it])+
            chi2*(m[i,it]*c2rr[i,it]+mr[i,it]*
            c2r[i,it]+(2/r[i])*m[i,it]*c2r[i,it])+
            f1(r[i]);
    c1t[i,it]=d1*(c1rr[i,it]+(2/r[i])*c1r[i,it])+
              a1*m[i,it]-b1*c1[i,it];
    c2t[i,it]=d2*(c2rr[i,it]+(2/r[i])*c2r[i,it])+
              a2*m[i,it]-b2*c2[i,it];
  }
 }
#
# Next t (it)
 }
#
# Plot PDE LHS (t derivatives)
#
# Plot mt
  par(mfrow=c(1,1));
  matplot(x=r,y=mt,type="l",xlab="r",
          ylab="mt(r,t)",xlim=c(0,r0),
          lty=1,main="mt(r,t)",lwd=2,
          col="black");
#
# Plot c1t
  par(mfrow=c(1,1));
  matplot(x=r,y=c1t,type="l",xlab="r",
```

```
          ylab="c1t(r,t)",xlim=c(0,r0),
          lty=1,main="c1t(r,t)",lwd=2,
          col="black");
#
# Plot c2t
  par(mfrow=c(1,1));
  matplot(x=r,y=c2t,type="l",xlab="r",
          ylab="c2t(r,t)",xlim=c(0,r0),
          lty=1,main="c2t(r,t)",lwd=2,
          col="black");
```

Listing 4.1: Main program with output of the t derivative vectors

Listing 4.1 is similar to Listing 3.2. Therefore, only the differences are indicated here.

- The ODE routine is now pde_1b rather than pde_1a, but the two routines are the same. The name change (from "a" to "b", with source("pde_1b.R")) is used to differentiate this second case with the analysis of the t derivative vectors.

```
#
# Access functions for numerical solution
  setwd("f:/neuro/chap4");
  source("pde_1b.R");source("f1.R");
  source("dss006.R");
  source("dss008.R");
```

dss008 is used for comparison with the dss006 solution in pde_b. The two solutions are essentially identical, thus inferring spatial convergence with nr=101. This comparison can be considered as a form of p refinement.[1]

---

[1]The error resulting from the truncation of the Taylor series on which the FD approximations of dss006 and dss008 are based, the truncation error,

- lsodes calls pde_1b.

```
#
# ODE integration
  out=lsodes(y=u0,times=tout,func=pde_1b,
      sparsetype ="sparseint",rtol=1e-6,
      atol=1e-6,maxord=5);
```

- Coding is added at the end for the calculation and display of the t derivative vectors. The first step is to declare the arrays used for the additional calculation and display of the t derivatives (the LHSs of eqs. (2.3) and (2.4)).

```
#
# LHS terms in PDEs
    mr=matrix(0,nrow=nr,ncol=nout);
   c1r=matrix(0,nrow=nr,ncol=nout);
   c2r=matrix(0,nrow=nr,ncol=nout);
   mrr=matrix(0,nrow=nr,ncol=nout);
  c1rr=matrix(0,nrow=nr,ncol=nout);
  c2rr=matrix(0,nrow=nr,ncol=nout);
    mt=matrix(0,nrow=nr,ncol=nout);
   c1t=matrix(0,nrow=nr,ncol=nout);
   c2t=matrix(0,nrow=nr,ncol=nout);
```

The arrays are 2D to accommodate variations in the solutions with respect to r and t, that is, $m(r,t), c_1(r,t), c_2(r,t)$. The first (row) dimension is for r (nr=101) and the second (column) dimension is for t (nout=6).

- The first derivatives, $\dfrac{\partial m}{\partial r}, \dfrac{\partial c_1}{\partial r}, \dfrac{\partial c_2}{\partial r}$, are computed by **dss006**.

---

is of the form $error = c\Delta x^p$ where Δx is the FD grid spacing. The FD approximations in **dss006** and **dss008** correspond to $p = 6$ and $p = 8$, respectively.

```
#
# mr,c1r,c2r
  for(it in 1:nout){
   mr[,it]=dss006(0,r0,nr, m[,it]);
   c1r[,it]=dss006(0,r0,nr,c1[,it]);
   c2r[,it]=dss006(0,r0,nr,c2[,it]);
```

Vectorization is used with , to specify all values of r, e.g., mr[,it]. Variations in t are specified with it, the index of the for.

- Homogeneous, Neumann BCs (2.5) are defined.

```
#
# BCs
   mr[1,it]=0;   mr[nr,it]=0;
   c1r[1,it]=0; c1r[nr,it]=0;
   c2r[1,it]=0; c2r[nr,it]=0;
```

Subscripts 1,nr specify the boundaries $r = 0, r_0$ for a particular value of t (with it).

- The PDEs, eqs. (2.3) and (2.4), are programmed in essentially the same way as in pde_1a (or pde_1b). The essential difference is the use of 2D arrays to account for both t (through it) and r (through i). The coding for eqs. (2.4) (for $r = 0$) follows.

```
#
# PDEs
  for(i in 1:nr){
    if(i==1){
      mt[i,it]=3*mu*mrr[i,it]-chi1*3*m[i,it]*
               c1rr[i,it]+chi2*3*m[i,it]*
               c2rr[i,it]+f1(r[i]);
      c1t[i,it]=3*d1*c1rr[i,it]+a1*m[i,it]-
               b1*c1[i,it];
```

```
    c2t[i,it]=3*d2*c2rr[i,it]+a2*m[i,it]-
           b2*c2[i,it];
   }
```

- The coding for eqs. (2.3) ($r \neq 0$) follows (from pde_1a).

```
   if(i>1){
     mt[i,it]=mu*(mrr[i,it]+(2/r[i])*mr[i,it])-
           chi1*(m[i,it]*c1rr[i,it]+mr[i,it]*
           c1r[i,it]+(2/r[i])*m[i,it]*c1r[i,it])+
           chi2*(m[i,it]*c2rr[i,it]+mr[i,it]*
           c2r[i,it]+(2/r[i])*m[i,it]*c2r[i,it])+
           f1(r[i]);
     c1t[i,it]=d1*(c1rr[i,it]+(2/r[i])*c1r[i,it])+
              a1*m[i,it]-b1*c1[i,it];
     c2t[i,it]=d2*(c2rr[i,it]+(2/r[i])*c2r[i,it])+
              a2*m[i,it]-b2*c2[i,it];
   }
  }
#
# Next t (it)
  }
```

An important detail of this coding for the derivatives in t is the use of the previously computed solution to eqs. (2.3) and (2.4) returned in out by lsodes. In other words, the derivatives in t (LHSs of eqs. (2.3) and (2.4)) can be computed using only the numerical PDE solution. The same is true for the computation of all of the RHS terms in the PDEs (they can be computed from the numerical PDE solution). The calculation and display of the LHS and RHS terms in the PDEs provides a valuable method for analyzing the PDEs, particularly the relative contributions of the various terms. This procedure is illustrated in the next main program and associated ODE routine.

The close correspondence between the t derivatives in eqs. (2.3) and (2.4) and the coding is clear, a principal advantage of the MOL.

- The three plots of $\dfrac{\partial m}{\partial t}$, $\dfrac{\partial c_1}{\partial t}$, $\dfrac{\partial c_2}{\partial t}$, against r with t as a parameter follows from calls to `matplot`.

```
#
# Plot PDE LHS (t derivatives)
#
# Plot mt
  par(mfrow=c(1,1));
  matplot(x=r,y=mt,type="l",xlab="r",
          ylab="mt(r,t)",xlim=c(0,r0),
          lty=1,main="mt(r,t)",lwd=2,
          col="black");
#
# Plot c1t
  par(mfrow=c(1,1));
  matplot(x=r,y=c1t,type="l",xlab="r",
          ylab="c1t(r,t)",xlim=c(0,r0),
          lty=1,main="c1t(r,t)",lwd=2,
          col="black");
#
# Plot c2t
  par(mfrow=c(1,1));
  matplot(x=r,y=c2t,type="l",xlab="r",
          ylab="c2t(r,t)",xlim=c(0,r0),
          lty=1,main="c2t(r,t)",lwd=2,
          col="black");
```

The three plots for **ncase=2** are in Figs. 4.3a, 4.3b and 4.3c (the plots for **ncase=1** are not presented since they just indicate that the derivatives in t remain at zero throughout the interval in t).

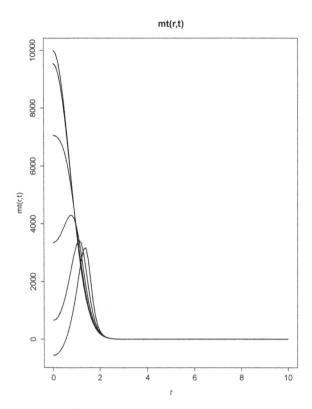

Figure 4.3a: $\partial m/\partial t$ against r with t as a parameter, `ncase=2`

Three important points can be mentioned about the t derivatives in Figs. 4.3.

- The derivatives in t (LHS terms in eqs. (2.3) and (2.4)) determine the solutions in Figs. 4.2. In other words, the programming of the t derivatives in `pde_1b` defines the PDE solutions.
- The transition of the t derivatives is smooth between $r = 0$ and $r > 0$, even with the switch in `pde_1b` (for `i=1` and `i > 1`).
- The t derivatives in Figs. 4.3 are complicated, and the question naturally occurs as to why they take this form. To answer this question, the contributions of the various RHS terms in eqs. (2.3) and (2.4) are examined with the following extension of the programming in Listing 4.1.

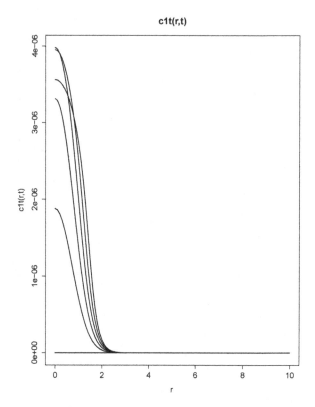

Figure 4.3b: $\partial c_1/\partial t$ against r with t as a parameter, `ncase=2`

(4.3) PDE Solution Vector with t Derivative Vectors and RHS Terms

The main program to compute the LHSs and RHS terms of eqs. (2.3) and (2.4) is listed next.

```
#
#  Three PDE chemotaxis model
#
# Access ODE integrator
  library("deSolve");
#
# Access functions for numerical solution
```

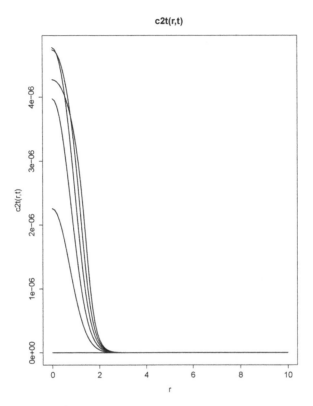

Figure 4.3c: $\partial c_2 / \partial t$ against r with t as a parameter, ncase=2

```
   setwd("f:/neuro/chap4");
   source("pde_1c.R");source("f1.R");
   source("dss006.R");
   source("dss008.R");
#
# Select case
   ncase=2;
#
# Parameters
   mu=3.0e-09;d1=1.0e-07;d2=1.0e-07;
   chi1=1;chi2=1;
   a1=1.0e-12;a2=1.2e-12;
```

```
  b1=5.0e-04;b2=5.0e-04;
  if(ncase==1){r1=0        ;r2=1;}
  if(ncase==2){r1=1.0e+04;r2=1;}
#
# Grid (in r)
  nr=101;r0=10;
  r=seq(from=0,to=r0,by=(r0-0)/(nr-1));
#
# Independent variable for ODE integration
  t0=0;tf=1.0e+03;nout=6;
  tout=seq(from=t0,to=tf,by=(tf-t0)/(nout-1));
#
# Initial condition
  u0=rep(0,3*nr);
  for(i in 1:nr){
    u0[i]     =0;
    u0[i+nr]   =0;
    u0[i+2*nr]=0;
  }
  ncall=0;
#
# ODE integration
  out=lsodes(y=u0,times=tout,func=pde_1c,
      sparsetype ="sparseint",rtol=1e-6,
      atol=1e-6,maxord=5);
  nrow(out)
  ncol(out)
#
# Arrays for plotting numerical solution
   m=matrix(0,nrow=nr,ncol=nout);
  c1=matrix(0,nrow=nr,ncol=nout);
  c2=matrix(0,nrow=nr,ncol=nout);

  for(it in 1:nout){
```

```
    tout[it]=tout[it]/60;
    for(ir in 1:nr){
       m[ir,it]=out[it,ir+1];
      c1[ir,it]=out[it,ir+1+nr];
      c2[ir,it]=out[it,ir+1+2*nr];
    }
  }
#
# Display numerical solution
  for(it in 1:nout){
    if((it==1)|(it==nout)){
      cat(sprintf("\n      t      r       m(r,t)
                    c1(r,t)      c2(r,t)\n"));
      for(ir in 1:nr){
        cat(sprintf("%6.2f%6.1f%12.3e%12.3e%12.3e\n",
        tout[it],r[ir],m[ir,it],c1[ir,it],c2[ir,it]));
      }
    }
  }
#
# Calls to ODE routine
  cat(sprintf("\n\n ncall = %5d\n\n",ncall));
#
# Plot PDE solutions
#
# m
  par(mfrow=c(1,1));
  matplot(x=r,y=m,type="l",xlab="r (cm)",
          ylab="m(r,t) (cells/cc)",
          xlim=c(0,r0),lty=1,main="m(r,t)",
          lwd=2,col="black");
#
# c1
  par(mfrow=c(1,1));
```

```
  matplot(x=r,y=c1,type="l",xlab="r (cm)",
          ylab="c1(r,t) (M = mols/liter)",
          xlim=c(0,r0),lty=1,main="c1(r,t)",
          lwd=2,col="black");
#
# c2
  par(mfrow=c(1,1));
  matplot(x=r,y=c2,type="l",xlab="r (cm)",
          ylab="c2(r,t) (M = mols/liter)",
          xlim=c(0,r0),lty=1,main="c2(r,t)",
          lwd=2,col="black");
#
# LHS terms in PDEs
    mr=matrix(0,nrow=nr,ncol=nout);
   c1r=matrix(0,nrow=nr,ncol=nout);
   c2r=matrix(0,nrow=nr,ncol=nout);
   mrr=matrix(0,nrow=nr,ncol=nout);
  c1rr=matrix(0,nrow=nr,ncol=nout);
  c2rr=matrix(0,nrow=nr,ncol=nout);
    mt=matrix(0,nrow=nr,ncol=nout);
   c1t=matrix(0,nrow=nr,ncol=nout);
   c2t=matrix(0,nrow=nr,ncol=nout);
#
# mr,c1r,c2r
  for(it in 1:nout){
   mr[,it]=dss006(0,r0,nr, m[,it]);
  c1r[,it]=dss006(0,r0,nr,c1[,it]);
  c2r[,it]=dss006(0,r0,nr,c2[,it]);
#
# BCs
   mr[1,it]=0;  mr[nr,it]=0;
  c1r[1,it]=0; c1r[nr,it]=0;
  c2r[1,it]=0; c2r[nr,it]=0;
#
```

```
# mrr,c1rr,c2rr
  mrr[,it]=dss006(0,r0,nr, mr[,it]);
  c1rr[,it]=dss006(0,r0,nr,c1r[,it]);
  c2rr[,it]=dss006(0,r0,nr,c2r[,it]);
#
# PDEs
  for(i in 1:nr){
    if(i==1){
      mt[i,it]=3*mu*mrr[i,it]-chi1*3*m[i,it]*
                  c1rr[i,it]+chi2*3*m[i,it]*
                  c2rr[i,it]+f1(r[i]);
      c1t[i,it]=3*d1*c1rr[i,it]+a1*m[i,it]-
                  b1*c1[i,it];
      c2t[i,it]=3*d2*c2rr[i,it]+a2*m[i,it]-
                  b2*c2[i,it];
    }
    if(i>1){
      mt[i,it]=mu*(mrr[i,it]+(2/r[i])*mr[i,it])-
                  chi1*(m[i,it]*c1rr[i,it]+mr[i,it]*
                  c1r[i,it]+(2/r[i])*m[i,it]*c1r[i,it])+
                  chi2*(m[i,it]*c2rr[i,it]+mr[i,it]*
                  c2r[i,it]+(2/r[i])*m[i,it]*c2r[i,it])+
                  f1(r[i]);
      c1t[i,it]=d1*(c1rr[i,it]+(2/r[i])*c1r[i,it])+
                  a1*m[i,it]-b1*c1[i,it];
      c2t[i,it]=d2*(c2rr[i,it]+(2/r[i])*c2r[i,it])+
                  a2*m[i,it]-b2*c2[i,it];
    }
  }
#
# Next t (it)
  }
#
# Plot PDE LHS (t derivatives)
```

```
#
# Plot mt
  par(mfrow=c(1,1));
  matplot(x=r,y=mt,type="l",xlab="r",
          ylab="mt(r,t)",xlim=c(0,r0),
          lty=1,main="mt(r,t)",lwd=2,
          col="black");
#
# Plot c1t
  par(mfrow=c(1,1));
  matplot(x=r,y=c1t,type="l",xlab="r",
          ylab="c1t(r,t)",xlim=c(0,r0),
          lty=1,main="c1t(r,t)",lwd=2,
          col="black");
#
# Plot c2t
  par(mfrow=c(1,1));
  matplot(x=r,y=c2t,type="l",xlab="r",
          ylab="c2t(r,t)",xlim=c(0,r0),
          lty=1,main="c2t(r,t)",lwd=2,
          col="black");
#
# RHS terms in PDEs
  term11=matrix(0,nrow=nr,ncol=nout);
  term12=matrix(0,nrow=nr,ncol=nout);
  term13=matrix(0,nrow=nr,ncol=nout);
  term14=matrix(0,nrow=nr,ncol=nout);
  term15=matrix(0,nrow=nr,ncol=nout);
  term16=matrix(0,nrow=nr,ncol=nout);
  term21=matrix(0,nrow=nr,ncol=nout);
  term22=matrix(0,nrow=nr,ncol=nout);
  term23=matrix(0,nrow=nr,ncol=nout);
  term31=matrix(0,nrow=nr,ncol=nout);
  term32=matrix(0,nrow=nr,ncol=nout);
```

```
   term33=matrix(0,nrow=nr,ncol=nout);
   rm=matrix(0,nrow=nr,ncol=nout);
#
# PDE RHS terms
   for(it in 1:nout){
   for(i in 1:nr){
     if(i==1){
       term11[i,it]=3*mu*mrr[i,it];
       term12[i,it]=-chi1*3*m[i,it]*c1rr[i,it];
       term13[i,it]= chi2*3*m[i,it]*c2rr[i,it];
       term14[i,it]=f1(r[i]);
       term15[i,it]=m[i,it]*c1r[i,it];
       term16[i,it]=m[i,it]*c2r[i,it];
       term21[i,it]=3*d1*c1rr[i,it];
       term22[i,it]=  a1*m[i,it];
       term23[i,it]=-b1*c1[i,it];
       term31[i,it]=3*d2*c2rr[i,it];
       term32[i,it]=  a2*m[i,it];
       term33[i,it]=-b2*c2[i,it];
     }
     if(i>1){
       term11[i,it]=mu*(mrr[i,it]+(2/r[i])*mr[i,it]);
       term12[i,it]=-chi1*(m[i,it]*c1rr[i,it]+
                   mr[i,it]*c1r[i,it]+(2/r[i])*
                    m[i,it]*c1r[i,it]);
       term13[i,it]= chi2*(m[i,it]*c2rr[i,it]+
                   mr[i,it]*c2r[i,it]+(2/r[i])*
                    m[i,it]*c2r[i,it]);
       term14[i,it]=f1(r[i]);
       term15[i,it]=m[i,it]*c1r[i,it];
       term16[i,it]=m[i,it]*c2r[i,it];
       term21[i,it]=d1*(c1rr[i,it]+(2/r[i])*
                   c1r[i,it]);
       term22[i,it]=  a1*m[i,it];
```

```
      term23[i,it]=-b1*c1[i,it];
      term31[i,it]=d2*(c2rr[i,it]+(2/r[i])*
                    c2r[i,it]);
      term32[i,it]= a2*m[i,it];
      term33[i,it]=-b2*c2[i,it];
    }
  }
#
# Next t (it)
  }
#
# Plot PDE RHS terms
#
# term11
  par(mfrow=c(1,1));
  matplot(x=r,y=term11,type="l",xlab="r",
          ylab="term11(r,t)",xlim=c(0,r0),
          lty=1,main="term11(r,t)",lwd=2,
          col="black");
#
# term12
  par(mfrow=c(1,1));
  matplot(x=r,y=term12,type="l",xlab="r",
          ylab="term12(r,t)",xlim=c(0,r0),
          lty=1,main="term12(r,t)",lwd=2,
          col="black");
#
# term13
  par(mfrow=c(1,1));
  matplot(x=r,y=term13,type="l",xlab="r",
          ylab="term13(r,t)",xlim=c(0,r0),
          lty=1,main="term13(r,t)",lwd=2,
          col="black");
#
```

```
# term14
  par(mfrow=c(1,1));
  matplot(x=r,y=term14,type="l",xlab="r",
          ylab="term14(r,t)",xlim=c(0,r0),
          lty=1,main="term14(r,t)",lwd=2,
          col="black");
#
# term15
  par(mfrow=c(1,1));
  matplot(x=r,y=term15,type="l",xlab="r",
          ylab="term15(r,t)",xlim=c(0,r0),
          lty=1,main="term15(r,t)",lwd=2,
          col="black");
#
# term16
  par(mfrow=c(1,1));
  matplot(x=r,y=term16,type="l",xlab="r",
          ylab="term16(r,t)",xlim=c(0,r0),
          lty=1,main="term16(r,t)",lwd=2,
          col="black");
#
# term21
  par(mfrow=c(1,1));
  matplot(x=r,y=term21,type="l",xlab="r",
          ylab="term21(r,t)",xlim=c(0,r0),
          lty=1,main="term21(r,t)",
          lwd=2,col="black");
#
# term22
  par(mfrow=c(1,1));
  matplot(x=r,y=term22,type="l",xlab="r",
          ylab="term22(r,t)",xlim=c(0,r0),
          lty=1,main="term22(r,t)",lwd=2,
          col="black");
```

```
#
# term23
  par(mfrow=c(1,1));
  matplot(x=r,y=term23,type="l",xlab="r",
          ylab="term23(r,t)",xlim=c(0,r0),
          lty=1,main="term23(r,t)",lwd=2,
          col="black");
#
# term31
  par(mfrow=c(1,1));
  matplot(x=r,y=term31,type="l",xlab="r",
          ylab="term31(r,t)",xlim=c(0,r0),
          lty=1,main="term31(r,t)",lwd=2,
          col="black");
#
# term32
  par(mfrow=c(1,1));
  matplot(x=r,y=term32,type="l",xlab="r",
          ylab="term32(r,t)",xlim=c(0,r0),
          lty=1,main="term32(r,t)",lwd=2,
          col="black");
#
# term33
  par(mfrow=c(1,1));
  matplot(x=r,y=term33,type="l",xlab="r",
          ylab="term33(r,t)",xlim=c(0,r0),
          lty=1,main="term33(r,t)",lwd=2,
          col="black");
```

Listing 4.2: Main program with output of the t derivative vectors and RHS terms of eqs. (2.3) and (2.4)

Listing 4.2 is similar to Listing 4.1. Therefore, only the differences are considered here.

- The ODE routine is now pde_1c rather than pde_1b, but the two routines are the same. The name change (from "b" to "c", with source("pde_1c.R")) is used to differentiate this third case with the analysis of the LHS t derivative vectors and the RHS terms.

```
#
# Access functions for numerical solution
  setwd("f:/neuro/chap4");
  source("pde_1c.R");source("f1.R");
  source("dss006.R");
  source("dss008.R");
```

- lsodes calls pde_1c.

```
#
# ODE integration
  out=lsodes(y=u0,times=tout,func=pde_1c,
      sparsetype ="sparseint",rtol=1e-6,
      atol=1e-6,maxord=5);
```

- Coding is added at the end for the calculation and display of the t derivative vectors, and the RHS terms of eqs. (2.3) and (2.4). Additional arrays are first declared for the RHS terms.

```
#
# RHS terms in PDEs
  term11=matrix(0,nrow=nr,ncol=nout);
  term12=matrix(0,nrow=nr,ncol=nout);
  term13=matrix(0,nrow=nr,ncol=nout);
  term14=matrix(0,nrow=nr,ncol=nout);
  term15=matrix(0,nrow=nr,ncol=nout);
  term16=matrix(0,nrow=nr,ncol=nout);
  term21=matrix(0,nrow=nr,ncol=nout);
  term22=matrix(0,nrow=nr,ncol=nout);
```

```
term23=matrix(0,nrow=nr,ncol=nout);
term31=matrix(0,nrow=nr,ncol=nout);
term32=matrix(0,nrow=nr,ncol=nout);
term33=matrix(0,nrow=nr,ncol=nout);
rm=matrix(0,nrow=nr,ncol=nout);
```

Again, the arrays are 2D to accommodate variations in the solutions with respect to r and t. The naming convention is `term-eqn-term`. For example, `term11` is for the first PDE (eq. (2.3a), (2.4a)) and the first term $\left(\mu\left(\dfrac{\partial^2 m}{\partial r^2}\right)\right)$. The first (row) dimension is for r (`nr=101`) and the second (column) dimension is for t (`nout=6`).

- Two `for`s step through t and r, with indices `it` and `i`, respectively. For $r = 0$ (`i=1`), the calculation of the RHS terms is

```
#
# PDE RHS terms
  for(it in 1:nout){
  for(i in 1:nr){
    if(i==1){
      term11[i,it]=3*mu*mrr[i,it];
      term12[i,it]=-chi1*3*m[i,it]*c1rr[i,it];
      term13[i,it]= chi2*3*m[i,it]*c2rr[i,it];
      term14[i,it]=f1(r[i]);
      term15[i,it]=m[i,it]*c1r[i,it];
      term16[i,it]=m[i,it]*c2r[i,it];
      term21[i,it]=3*d1*c1rr[i,it];
      term22[i,it]=  a1*m[i,it];
      term23[i,it]=-b1*c1[i,it];
      term31[i,it]=3*d2*c2rr[i,it];
      term32[i,it]=  a2*m[i,it];
      term33[i,it]=-b2*c2[i,it];
    }
```

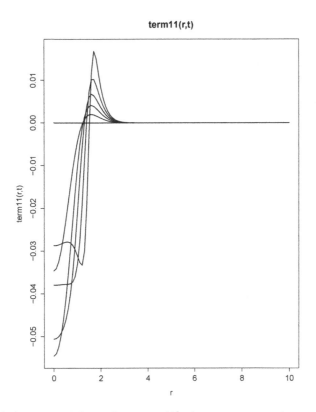

term11(r,t)

Figure 4.4a: `term11` against r with t as a parameter, `ncase=2`

The spatial derivatives in r calculated previously are used to calculate the RHS terms, e.g., `mrr[i,it]` in `term11[i,it]=3*mu*mrr[i,it]`. The first PDE includes two additional terms, `term15[i,it]=m[i,it]*c1r[i,it]`, `term16[i,it]=m[i,it]*c2r[i,it]`, to give an indication of the flux magntitudes from chemotaxis.

- For $r \neq 0$ (`i>1`), the calculation of the RHS terms is

```
if(i>1){
    term11[i,it]=mu*(mrr[i,it]+(2/r[i])*mr[i,it]);
    term12[i,it]=-chi1*(m[i,it]*c1rr[i,it]+
```

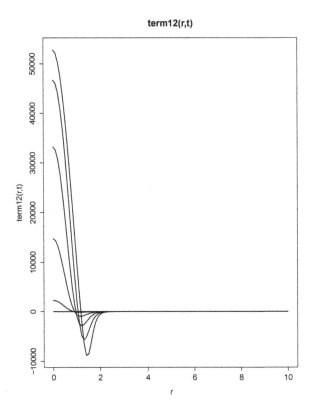

term12(r,t)

Figure 4.4b: `term12` against r with t as a parameter, `ncase=2`

```
                    mr[i,it]*c1r[i,it]+(2/r[i])*
                    m[i,it]*c1r[i,it]);
    term13[i,it]= chi2*(m[i,it]*c2rr[i,it]+
                    mr[i,it]*c2r[i,it]+(2/r[i])*
                    m[i,it]*c2r[i,it]);
    term14[i,it]=f1(r[i]);
    term15[i,it]=m[i,it]*c1r[i,it];
    term16[i,it]=m[i,it]*c2r[i,it];
    term21[i,it]=d1*(c1rr[i,it]+(2/r[i])*
                    c1r[i,it]);
    term22[i,it]=  a1*m[i,it];
```

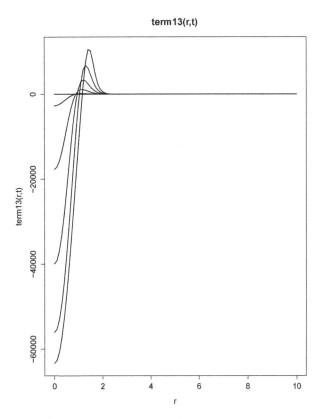

term13(r,t)

Figure 4.4c: `term13` against r with t as a parameter, `ncase=2`

```
    term23[i,it]=-b1*c1[i,it];
    term31[i,it]=d2*(c2rr[i,it]+(2/r[i])*
                c2r[i,it]);
    term32[i,it]=  a2*m[i,it];
    term33[i,it]=-b2*c2[i,it];
  }
 }
#
# Next t (it)
  }
```

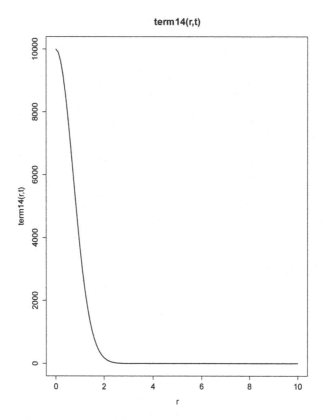

Figure 4.4d: `term14` against r with t as a parameter, `ncase=2`

- The RHS terms of the PDEs, eqs. (2.3), (2.4), are plotted with one plot on a page.[2]

```
#
# Plot PDE RHS terms
#
# term11
```

---

[2]Several attempts at placing more than one plot on a page, e.g., using `par(mfrow=c(1,2))`, `par(mfrow=c(2,1))`, `par(mfrow=c(2,2))`, produced poorly scaled plots that were difficult to interpret and compare. Therefore, one plot on a page was selected, even though this produced a total of 12 plots (for `term11` to `term33`).

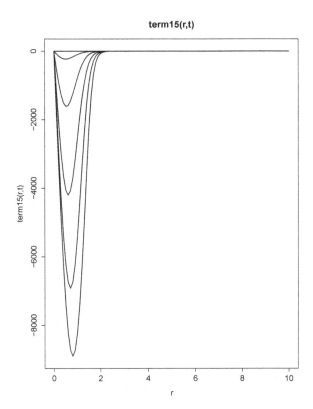

Figure 4.4e: `term15` against r with t as a parameter, `ncase=2`

```
  par(mfrow=c(1,1));
  matplot(x=r,y=term11,type="l",xlab="r",
          ylab="term11(r,t)",xlim=c(0,r0),
          lty=1,main="term11(r,t)",lwd=2,
          col="black");
#
# term12
  par(mfrow=c(1,1));
  matplot(x=r,y=term12,type="l",xlab="r",
          ylab="term12(r,t)",xlim=c(0,r0),
          lty=1,main="term12(r,t)",lwd=2,
          col="black");
```

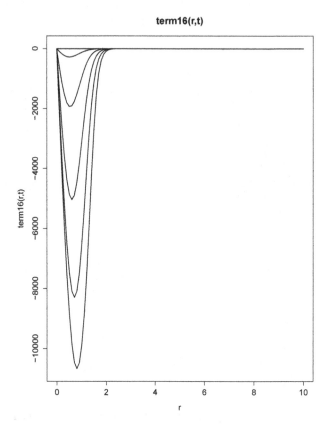

Figure 4.4f: `term16` against r with t as a parameter, `ncase=2`

```
#
# term13
  par(mfrow=c(1,1));
  matplot(x=r,y=term13,type="l",xlab="r",
          ylab="term13(r,t)",xlim=c(0,r0),
          lty=1,main="term13(r,t)",lwd=2,
          col="black");
#
# term14
  par(mfrow=c(1,1));
  matplot(x=r,y=term14,type="l",xlab="r",
```

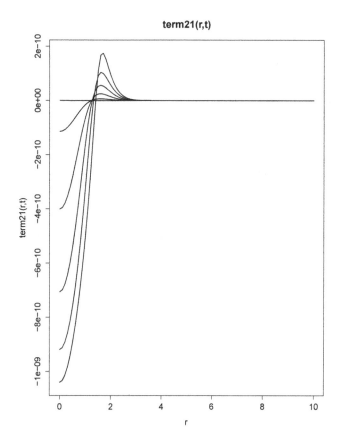

Figure 4.5a: `term21` against r with t as a parameter, `ncase=2`

```
        ylab="term14(r,t)",xlim=c(0,r0),
        lty=1,main="term14(r,t)",lwd=2,
        col="black");
#
# term15
  par(mfrow=c(1,1));
  matplot(x=r,y=term15,type="l",xlab="r",
        ylab="term15(r,t)",xlim=c(0,r0),
        lty=1,main="term15(r,t)",lwd=2,
        col="black");
```

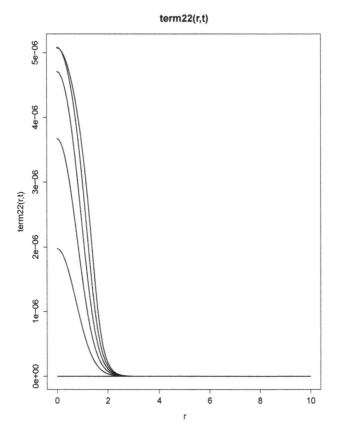

Figure 4.5b: `term22` against r with t as a parameter, `ncase=2`

```
#
# term16
  par(mfrow=c(1,1));
  matplot(x=r,y=term16,type="l",xlab="r",
          ylab="term16(r,t)",xlim=c(0,r0),
          lty=1,main="term16(r,t)",lwd=2,
          col="black");
#
# term21
  par(mfrow=c(1,1));
```

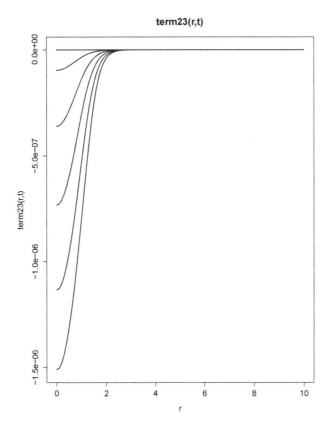

Figure 4.5c: **term23** against r with t as a parameter, **ncase=2**

```
    matplot(x=r,y=term21,type="l",xlab="r",
            ylab="term21(r,t)",xlim=c(0,r0),
            lty=1,main="term21(r,t)",
            lwd=2,col="black");
  #
  # term22
    par(mfrow=c(1,1));
    matplot(x=r,y=term22,type="l",xlab="r",
            ylab="term22(r,t)",xlim=c(0,r0),
            lty=1,main="term22(r,t)",lwd=2,
            col="black");
```

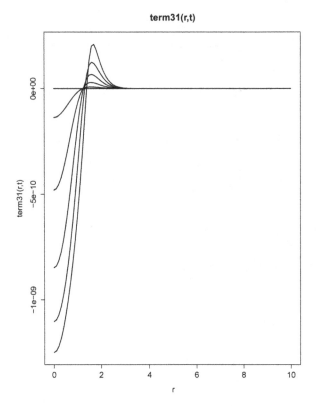

Figure 4.6a: `term31` against r with t as a parameter, `ncase=2`

```
#
# term23
  par(mfrow=c(1,1));
  matplot(x=r,y=term23,type="l",xlab="r",
          ylab="term23(r,t)",xlim=c(0,r0),
          lty=1,main="term23(r,t)",lwd=2,
          col="black");
#
# term31
  par(mfrow=c(1,1));
```

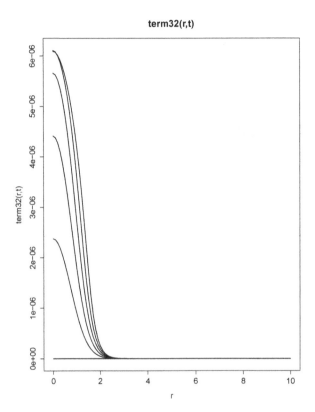

Figure 4.6b: `term32` against r with t as a parameter, `ncase=2`

```
matplot(x=r,y=term31,type="l",xlab="r",
        ylab="term31(r,t)",xlim=c(0,r0),
        lty=1,main="term31(r,t)",lwd=2,
        col="black");
#
# term32
par(mfrow=c(1,1));
matplot(x=r,y=term32,type="l",xlab="r",
        ylab="term32(r,t)",xlim=c(0,r0),
        lty=1,main="term32(r,t)",lwd=2,
        col="black");
#
```

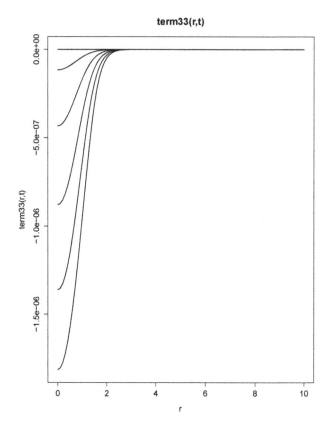

Figure 4.6c: `term33` against r with t as a parameter, `ncase=2`

```
# term33
  par(mfrow=c(1,1));
  matplot(x=r,y=term33,type="l",xlab="r",
          ylab="term33(r,t)",xlim=c(0,r0),
          lty=1,main="term33(r,t)",lwd=2,
          col="black");
```

The plots that result from the preceding code are in Figs. 4.4a,b,c,d,e,f for the first PDE (eqs. (2.3a) and (2.4a)), Figs. 4.5a,b,c for the second PDE (eqs. (2.3b) and (2.4b)), and Figs. 4.6a,b,c for the third PDE (eqs. (2.3c) and (2.4c)).

These plots give a detailed picture of the relative contributions of ordinary diffusion, chemotaxis diffusion (movement), chemical reactions and cell production. In particular, some of the terms might be neglibly small compared to others in a given PDE, and could therefore be dropped, or the parameters (constants) that define the small terms might be increased.

For example, the approximate intervals for `term11` to `term16` in eqs. (2.3a) and (2.4a) are

Figure	Interval
Fig. 4.4a	$-0.05 \leq$ `term11` ≤ 0
Fig. 4.4b	$-10000 \leq$ `term12` ≤ 50000
Fig. 4.4c	$-60000 \leq$ `term13` ≤ 0
Fig. 4.4d	$0 \leq$ `term14` ≤ 10000
Fig. 4.4e	$-8000 \leq$ `term15` ≤ 0
Fig. 4.4f	$-10000 \leq$ `term16` ≤ 0

Table 4.2: Intervals for `term11` to `term16` in eqs. (2.3a), (2.4a).

We can note the following details about the intervals in Table 4.2.

- `term11`, which is a measure of the Fickian diffusion of cells in eqs. (2.3a), (2.4a), is negligibly small (relative to the other terms). Therefore, either this term could be dropped from the model, or the diffusivity μ could be increased to give a significant contribution to the cell derivative $\dfrac{\partial m}{\partial t}$.
- `term12`, which is a measure of the effect of the attractant, is generally large and positive, i.e., it increases $\dfrac{\partial m}{\partial t}$, as expected.

- **term13**, which is a measure of the effect of the repellant, is large and negative, i.e., it decreases $\frac{\partial m}{\partial t}$, as expected.

- **term14**, which is the cell volumetric source term, does not change with t since the Gaussian $f(m, c_1, c_2) = r_1 e^{-r_2 r^2}$ is a function of r only. Also, it is positive so that it increases $\frac{\partial m}{\partial t}$ with t, and it is this term that moves the model away from the homogeneous ICs. Since this term does not change with t, it continually increases the t derivatives $\frac{\partial m}{\partial t}, \frac{\partial c_1}{\partial t}, \frac{\partial c_2}{\partial t}$, as reflected in Figs. 4.3a, 4.3b and 4.3c and thereby continually increases $m(r, t), c_1(r, t), c_2(r, t)$ as reflected in Figs. 4.2a, 4.2b, 4.2c.

- The various terms generally start out at zero because of the homogeneous BCs, and homogeneous Neumann BCs (eqs. (4.5)). Then they move away from zero in response to the source term **term14** (which could, for example, represent local inflammation that initiates AD). The time scale for the developing $m(r, t), c_1(r, t), c_2(r, t)$ is of the order of minutes, which is much shorter (faster) than the time scale for AD, which is of the order of 5–20 years. This difference in time scales reflects the relatively slow progression of AD, and the fact that the solutions of eqs. (2.3) and (2.4) are essentially at equilibrium or steady state as the AD develops.

- **term15** and **term16**, which are a measure of the chemotaxis of cells due to the attractant and repellent, respectively, are negative. However, they are opposite in sign in eqs. (2.3a) and (2.4a) due to multiplication by $-\chi_1$ and χ_2 (with $\chi_1 > 0$, $\chi_2 > 0$), respectively, and therefore have opposite effects on $\frac{\partial m}{\partial t}$ as expected (**term15** tends to increase $\frac{\partial m}{\partial t}$ while **term16** tends to decrease $\frac{\partial m}{\partial t}$).

- **term12** to **term14** are of approximately equal magnitude, but their sum is positive so that $\frac{\partial m}{\partial t}$ is positive (Fig. 4.3a) and therefore $m(r, t)$ increases (Fig. 4.2a) with t (as expected for increasing cell density).

A similar analysis can be applied to term21 to term23 for eqs. (2.3b) and (2.4b), and to term31 to term33 for eqs. (2.3c) and (2.4c). With the insights provided by this detailed analysis of the LHS and RHS terms in the PDEs, understanding of the PDEs and their numerical solutions is enhanced. Ideally, the model development can be continued until a form is achieved that provides the desired insights into the chemotaxis system, possibly including the interpretation of experimental observations and data.

For all of the solutions discussed, homogeneous, Neumann BCs (2.5) are satisfied (Figs. 4.2a, 4.2b, 4.2c). Also, the homogeneous ICs are compatible with these BCs. If nonzero ICs are used instead, they may not be compatible with the BCs which would introduce a discontinuity at the boundaries $r = 0, r_0$ at $t = 0$. The response of the numerical solution to this discontinuity could be observed in the numerical and graphical output.

Some extensions of the model and the accuracy of the solutions are discussed in the following chapters.

Reference

[1] Luca, M., A. Chavez-Ross, L. Edelstein-Keshet and A. Mogilner (2003), Chemotactic signaling, microglia, and Alzheimer's disease senile plaques: Is there a connection?, *Bulletin of Mathematical Biology*, **65**, 693–730

Chapter 5

Model Variants

A principal advantage of having a computer-based model for a problem system of interest is the opportunity to experiment with the model and observe the resulting solutions. This experimentation can take several forms:

- Variation of the model parameters (usually constants) to observe their effect on the solution. In particular, identifying parameters which are sensitive in the sense that small changes in their values produces significant changes in the solution is important. i.e., the values of those parameters must be established as accurately and reliably as possible.
- Variation of the terms in the model equation. This can include changing the form of the terms, e.g., linear to nonlinear, adding new terms to reflect phenomena that might be considered important in the analysis and deletion of terms which have little or no effect on the solution, or which produce features of the model output that are clearly unrealistic physically.
- Addition or deletion of equations to extend the analysis.
- Formulation of a new model using the methodology implemented in the current model, e.g., the same numerical algorithms and coding (as a prototype).

This list indicates that the formulation of a computer-based model for a new application is generally a trial-and-error procedure, even when a model has been reported in the literature.

For the latter, coding of the reported model may be the first step, and this will almost certainly require some trial-and-error to produce a series of functional routines.

In this chapter, we consider some of these possibilities, particularly variations in the PDE model of eqs. (2.3) to (2.6). The detailed examination of the contribution of individual terms in the PDEs, as illustrated in Chapter 4 (with graphical output in Figs. 4.4a to 4.4f), can be particularly helpful in the evolutionary process of building a model.

(5.1) Effect of the Attractor/Repellent

The solution of eqs. (2.3) to (2.6), as implemented in Listings 3.1 and 3.3, corresponds to a near offsetting of the chemotaxis attraction and repulsion terms of eq. (2.3a)

$$-\chi_1 \left(m\frac{\partial^2 c_1}{\partial r^2} + \frac{\partial m}{\partial r}\frac{\partial c_1}{\partial r} + \frac{2}{r}m\frac{\partial c_1}{\partial r} \right)$$

$$+\chi_2 \left(m\frac{\partial^2 c_2}{\partial r^2} + \frac{\partial m}{\partial r}\frac{\partial c_2}{\partial r} + \frac{2}{r}m\frac{\partial c_2}{\partial r} \right)$$

and eq. (2.4a)

$$-\chi_1 \left(3m\frac{\partial^2 c_1}{\partial r^2} + \frac{\partial m}{\partial r}\frac{\partial c_1}{\partial r} \right)$$

$$+\chi_2 \left(3m\frac{\partial^2 c_2}{\partial r^2} + \frac{\partial m}{\partial r}\frac{\partial c_2}{\partial r} \right)$$

Note that the two terms with χ_1 and χ_2 are opposite in sign and nearly of the same magnitude. These terms are plotted as term12, term13 in Figs. 4.4b, 4.4c, respectively, which illustrate their offsetting magnitudes.

The individual χ_1 and χ_2 terms can be investigated to demonstrate the effect of attraction and repulsion, respectively. In particular, the case of no repulsion is considered next with $\chi_2 = 0$ (everything else in Listing 3.2 is unchanged, and the ODE/MOL routine is taken from Listing 3.1).

```
#
# Parameters
  mu=3.0e-09;d1=1.0e-07;d2=1.0e-07;
  chi1=1;chi2=0;
  a1=1.0e-12;a2=1.2e-12;
  b1=5.0e-04;b2=5.0e-04;
  r1=1.0e+04;r2=1;
```

With only attraction (chi1=1;chi2=0;), the ODE integration fails (lsodes gives no output other than an error message indicating an excessively small integration step).

The selection of $\chi_1 = 1$, $\chi_2 = 0$ demonstrates the possibilities for varying parameters, and the unexpected effect that might result (particularly for parameters in nonlinear terms, which can demonstrate unexpected parameter sensitivity).

The most challenging (sometimes intractable) problem in PDE/MOL analysis is usually an ODE library integrator failure. In particular, since the detailed operation of the integrator is not readily available to the analyst, for example, the internal h and p refinement to attempt to meet the integration error tolerances, the cause of an integrator failure may not be clear.

A resolution might be possible with some experimentation, e.g., variation in the integrator error tolerances and other parameters, although for the present case, using absolute and relative error tolerances of 1.0e-04 or 1.0e-12 (1.0e-06 is the default) in the call to lsodes did not resolve the problem of the integration failure.

Another approach to resolving an integration failure is the use of an alternate ODE integrator such as the in-line (explicit) programming of an integration algorithm, e.g., the Euler or Runge Kutta method, that can be followed in detail during execution. This approach is demonstrated with the following main program (in place of Listing 3.2).

```
#
#  Three PDE chemotaxis model
#
# Access ODE integrator
  library("deSolve");
#
# Access functions for numerical solution
  setwd("f:/neuro/chap5");
  source("pde_1b.R");source("f1.R");
  source("dss006.R");
  source("dss008.R");
#
# Select case
  ncase=1;
#
# Parameters
  mu=3.0e-09;d1=1.0e-07;d2=1.0e-07;
  chi1=1;chi2=0;
  a1=1.0e-12;a2=1.2e-12;
  b1=5.0e-04;b2=5.0e-04;
  r1=1.0e+04;r2=1;
#
# Grid (in r)
  nr=101;r0=10;
  r=seq(from=0,to=r0,by=(r0-0)/(nr-1));
#
# Independent variable for ODE integration
  t0=0;tf=1.0e+03;nout=6;
  tout=seq(from=t0,to=tf,by=(tf-t0)/(nout-1));
#
# Initial condition
  u0=rep(0,3*nr);
  for(i in 1:nr){
    u0[i]     =0;
    u0[i+nr]  =0;
```

```
    u0[i+2*nr]=0;
  }
  ncall=0;
#
# ODE integration
# out=lsodes(y=u0,times=tout,func=pde_1a,
#     sparsetype ="sparseint",rtol=1e-6,
#     atol=1e-6,maxord=5);
# nrow(out)
# ncol(out)
#
# ODE integration
  m=matrix(0,nrow=nr,ncol=nout);
  c1=matrix(0,nrow=nr,ncol=nout);
  c2=matrix(0,nrow=nr,ncol=nout);
  u=u0;t=t0;h=2;
  for(it in 1:nout){
    for(ir in 1:nr){
       m[ir,it]=u[ir];
       c1[ir,it]=u[ir+nr];
       c2[ir,it]=u[ir+2*nr];
    }
    for(ir in 1:100){
      derv=pde_1b(t,u,parm);
      u=u+derv*h;
      t=t+h;
    }
  }
#
# Display numerical solution
  for(it in 1:nout){
    if((it==1)|(it==nout)){
      cat(sprintf("\n     t      r       m(r,t)
              c1(r,t)      c2(r,t)\n"));
```

```
      for(ir in 1:nr){
      cat(sprintf("%6.2f%6.1f%12.3e%12.3e%12.3e\n",
                  tout[it],r[ir],m[ir,it],c1[ir,it],
                  c2[ir,it]));
      }
    }
  }
#
# Calls to ODE routine
  cat(sprintf("\n\n ncall = %5d\n\n",ncall));
#
# Plot PDE solutions
#
# m
  par(mfrow=c(1,1));
  matplot(x=r,y=m,type="l",xlab="r (cm)",
          ylab="m(r,t) (cells/cc)",
          xlim=c(0,r0),lty=1,main="m(r,t)",
          lwd=2,col="black");
#
# c1
  par(mfrow=c(1,1));
  matplot(x=r,y=c1,type="l",xlab="r (cm)",
          ylab="c1(r,t) (M = mols/liter)",
          xlim=c(0,r0),lty=1,main="c1(r,t)",
          lwd=2,col="black");
#
# c2
  par(mfrow=c(1,1));
  matplot(x=r,y=c2,type="l",xlab="r (cm)",
          ylab="c2(r,t) (M = mols/liter)",
          xlim=c(0,r0),lty=1,main="c2(r,t)",
          lwd=2,col="black");
```

Listing 5.1: Main program with an in-line Euler integrator

We can note the following details about Listing 5.1.

- The ODE library deSolve and the subordinate routines pde_1b, f1, dss006, dss008 are specified.

```
#
#  Three PDE chemotaxis model
#
# Access ODE integrator
  library("deSolve");
#
# Access functions for numerical solution
  setwd("f:/neuro/chap5");
  source("pde_1b.R");source("f1.R");
  source("dss006.R");
  source("dss008.R");
```

Note that the ODE/MOL routine pde_1b is given the designation "b" so that it is specific to this particular MOL formulation.

- The parameters for no repulsion (chi2=0) are used.

```
#
# Parameters
  mu=3.0e-09;d1=1.0e-07;d2=1.0e-07;
  chi1=1;chi2=0;
  a1=1.0e-12;a2=1.2e-12;
  b1=5.0e-04;b2=5.0e-04;
  r1=1.0e+04;r2=1;
```

- The spatial grid is defined, consisting of 101 points over the interval $r = 0 \leq r \leq r = r_0 = 10$ using the seq operator.

```
#
# Grid (in r)
  nr=101;r0=10;
  r=seq(from=0,to=r0,by=(r0-0)/(nr-1));
```

- The interval in t is defined as $0 \leq t \leq 10^3$ with 6 output points. Thus, the solution is displayed at $t = 0, 200, \ldots,$ 1000 sec.

```
#
# Independent variable for ODE integration
  t0=0;tf=1.0e+03;nout=6;
  tout=seq(from=t0,to=tf,by=(tf-t0)/(nout-1));
```

- The ICs (2.6) are again homogeneous (zero) so that the solution is the response to the Gaussian volumetric source term in eqs. (2.3a) and (2.4a). The total number of ICs corresponds to the number of ODEs (303).

```
#
# Initial condition
  u0=rep(0,3*nr);
  for(i in 1:nr){
    u0[i]     =0;
    u0[i+nr]  =0;
    u0[i+2*nr]=0;
  }
  ncall=0;
```

- lsodes is replaced with the in-line Euler integrator.

```
#
# ODE integration
# out=lsodes(y=u0,times=tout,func=pde_1a,
#       sparsetype ="sparseint",rtol=1e-6,
#       atol=1e-6,maxord=5);
# nrow(out)
# ncol(out)
#
# ODE integration
```

```
  m=matrix(0,nrow=nr,ncol=nout);
c1=matrix(0,nrow=nr,ncol=nout);
c2=matrix(0,nrow=nr,ncol=nout);
u=u0;t=t0;h=2;
for(it in 1:nout){
  for(ir in 1:nr){
     m[ir,it]=u[ir];
     c1[ir,it]=u[ir+nr];
     c2[ir,it]=u[ir+2*nr];
  }
  if(it==nout)break;
  for(i in 1:100){
    derv=pde_1b(t,u,parm);
    u=u+derv*h;
    t=t+h;
  }
}
```

The Euler integrator requires some additional explanation.

– Arrays are defined with the **matrix** utility for the numerical solution of eqs. (2.3) and (2.4), that is, for $m(r,t)$, $c_1(r,t)$, $c_2(r,t)$.

```
#
# ODE integration
   m=matrix(0,nrow=nr,ncol=nout);
c1=matrix(0,nrow=nr,ncol=nout);
c2=matrix(0,nrow=nr,ncol=nout);
```

The dimensions of these arrays are **nr x nout = 101 x 6** or **606** elements to define the PDE dependent variables in space r (first dimension) and time t (second dimension).
– The solution vector (**3 x 101 = 303** dependent variables of the MOL/ODEs) is equated to the IC vector **u0**. t is initialized to **t0 = 0**. The Euler integration step is set

to h=2 (a constant throughout the integration of the 303 ODEs).

```
u=u0;t=t0;h=2;
```

– Two nested **fors** step through the numerical solution in t (index it) and r (index ir).

```
for(it in 1:nout){
   for(ir in 1:nr){
      m[ir,it]=u[ir];
      c1[ir,it]=u[ir+nr];
      c2[ir,it]=u[ir+2*nr];
   }
```

The solution from the Euler integrator, u, is placed in the three arrays m,c1,c2. Note the use of the subscripting in u for the three PDEs.

– 100 Euler steps are taken along the solution, each of length h=2.

```
   if(it==nout)break;
   for(i in 1:100){
      derv=pde_1b(t,u,parm);
      u=u+derv*h;
      t=t+h;
   }
}
```

Briefly, the Euler method is

$$u_{i+1} = u_i + \frac{du_i}{dt}h$$

where i is the index for an Euler step along the solution. The derivative vector $\dfrac{du_i}{dt}$ is computed by a call to the ODE routine **pde_1b** discussed subsequently.[1] **pde_1b** has as input arguments **t**, the current value of t, **u**, the current solution vector, and **parm** for passing parameters to **pde_1b** which is unused.

- The ODE/MOL routine **pde_1b** is taken directly from Listing 3.1, but one important change is then required. The derivative vector from Listing 3.1 is returned to the ODE integrator **lsodes** as a list, which is a requirement of the ODE integrators in the library **deSolve**.

```
#
# Return derivative vector
  return(list(c(ut)));
  }
```

However, the calculation of the Euler solution u=u+derv*h requires that the derivative vector **derv** from **pde_1b** is

---

[1]The derivative vector is evaluated at i so this is the *explicit* Euler method. The derivative vector evaluated at $i + 1$ is the *implicit* Euler method. The essential difference is in the stability of the two methods. The explicit Euler method is conditionally stable with a limit on the integration step h (for stability), while the implicit Euler method is *unconditionally stable* (with no stability limit on h). The additional computation to achieve the greater stability of the implicit Euler method follows from the solution of systems of equations (nonlinear if the ODE/MOL system is nonlinear). This additional computation is justified if the ODE system is stiff since a larger integration step is possible (exceeding the stable step of the explicit Euler method).

Both methods are first order accurate, that is, the truncation error is $O(h)$. Higher order methods such as the classical fourth order Runge Kutta (RK) could be programmed in essentially the same way as the Euler method, but this improved accuracy is apparently not required (the Euler method, which is *the* first order RK, produced the same numerical solution as **lsodes**).

The use of the Euler method and **lsodes** can be considered as a form of p refinement. The term p refinement pertains to the truncation error of an algorithm of the form $O(\Delta h^p)$ where Δh is the time grid interval. For the Euler and classical RK methods, $p = 1, 4$, respectively.

numerical (u and h are numerical). Thus, the derivative vector computed in pde_1b is returned as a numerical vector.

```
#
# Return derivative vector
  return(c(ut));
```

Everything else in Listing 3.1 is unchanged.
- The statement for the Euler step, u=u+derv*h, uses vectorization, i.e., u and derv are 303-vectors, and h is 1-vector (a scalar). t is advanced along the solution with t=t+h.
- After 100 Euler steps, the solution has advanced by $100(2) = 200$ in t so that the solution has moved through one output interval and can therefore be stored in arrays m,c1,c2 for subsequent numerical and graphical output (the next value of it).

- Selected numerical output at $t = 0, 1000$ (it=1,nout=6) is displayed.

```
#
# Display numerical solution
  for(it in 1:nout){
    if((it==1)|(it==nout)){
      cat(sprintf("\n      t       r       m(r,t)
                c1(r,t)      c2(r,t)\n"));
      for(ir in 1:nr){
      cat(sprintf("%6.2f%6.1f%12.3e%12.3e%12.3e\n",
                tout[it]/60,r[ir],m[ir,it],
                c1[ir,it],c2[ir,it]));
      }
    }
  }
```

The three PDE dependent variables, $m(r,t), c_1(r,t), c_2(r,t)$, are displayed as a function of r and t (subscripts ir,it). Note the conversion of t from sec to min through the division by 60.

- The counter for the calls to pde_1b is displayed as a measure of the computational effort to produce the numerical solution of eqs. (2.3) to (2.6).

```
#
# Calls to ODE routine
  cat(sprintf("\n\n ncall = %5d\n\n",ncall));
```

- The numerical solutions are displayed graphically with the matplot utility.

```
#
# Plot PDE solutions
#
# m
  par(mfrow=c(1,1));
  matplot(x=r,y=m,type="l",xlab="r (cm)",
          ylab="m(r,t) (cells/cc)",
          xlim=c(0,r0),lty=1,main="m(r,t)",
          lwd=2,col="black");
#
# c1
  par(mfrow=c(1,1));
  matplot(x=r,y=c1,type="l",xlab="r (cm)",
          ylab="c1(r,t) (M)",xlim=c(0,r0),
          lty=1,main="c1(r,t)",lwd=2,
          col="black");
#
# c2
  par(mfrow=c(1,1));
  matplot(x=r,y=c2,type="l",xlab="r (cm)",
```

```
ylab="c2(r,t) (M)",xlim=c(0,r0),
lty=1,main="c2(r,t)",lwd=2,
col="black");
```

The solutions are plotted against r with t as a parameter.

The integration by lsodes terminated with no output other than an error message. The integration by the Euler method proceeds (through 200 steps for each output interval as discussed previously) until the calculation becomes unstable so that the computer arithmetic fails (e.g., leads to NaN results as occurs for $t > 400$).

The graphical output in Figs. 5.1a, 5.1b and 5.1c suggests that the solution is increasing without bound (is unstable) so that the solution proceeds only through the values $t = 0, 200,$ 400.[2] The reason for this result basically is related to the use of the attraction term in eqs. (2.3a) and (2.4a) with no repellent term (with $\chi_1 = 1$, $\chi_2 = 0$). A discussion of why the MOL approximation did not produce a complete stable solution (to $t = 1000$) is given in appendix A1.

The output in Figs. 5.1a, 5.1b, 5.1c could be further explained by plotting the derivatives $\dfrac{dm}{dt}, \dfrac{dc_1}{dt}, \dfrac{dc_2}{dt}$ computed in pde_1c. Additionally, the attractant terms

$$-\chi_1 \left(m\frac{\partial^2 c_1}{\partial r^2} + \frac{\partial m}{\partial r}\frac{\partial c_1}{\partial r} + \frac{2}{r}m\frac{\partial c_1}{\partial r} \right)$$

$$-\chi_1 \left(3m\frac{\partial^2 c_1}{\partial r^2} + \frac{\partial m}{\partial r}\frac{\partial c_1}{\partial r} \right)$$

in eqs. (2.3a), (2.4a) could be computed and plotted.

A series of runs with the code indicated that the repellent terms should be at least as large as the attractant terms for

---

[2]Figs. 5.1a, 5.1b, 5.1c were produced by executing the Euler integrator over the stable interval $0 \le t \le 400$, with graphical output at $t = 0, 200, 400$ (the three curves in Figs. 5.1a, 5.1b, 5.1c, starting from the homogeneous ICs).

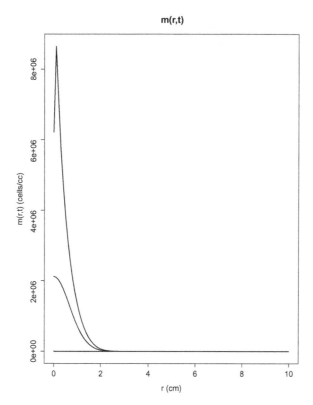

Figure 5.1a: $m(r,t)$ against r with t as a parameter, `chi2=0`

a stable, smooth solution, that is, $\chi_2 \geq \chi_1$, all else being equal (see appendix A1 for further discussion). This requirement might be relaxed, for example, if the Fickian diffusion in eqs. (2.3a), (2.4a) is increased, that is, μ, D_1, D_2 are increased (to smooth and spread the solutions in r).

In general, the balancing of attractant and repellent dynamics appears to be necessary to produce a stable solution.[3] In other words, the observed instability (blow-up) appears to be

---

[3]The solutions from `lsodes` (Chapter 4) and from the Euler integrator for $\chi_1 = 1$, $\chi_2 = 1$ are stable and smooth, and agree to 4-5 figures. These solutions generally increase monotonically with t from the source term of function `f1` used in eq. (2.3a), (2.4a) which is constant in t (Listing 3.1).

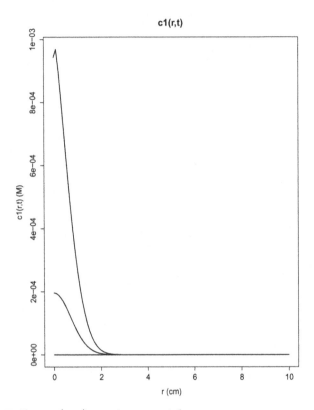

Figure 5.1b: $c_1(r, t)$ against r with t as a parameter, `chi2=0`

a feature of the model, and not of the numerical PDE (MOL) algorithm. This conclusion is substantiated in [1].[4]

The Euler method provides step-by-step insight into the details of the model dynamics and numerical integration that might not be available from a library integrator. Detailed output at each integration step would elucidate these details, e.g., how an instability develops with increasing t.

---

[4]The analysis in [1] is for eqs. (2.3) and (2.4) with no source term ($f(m, c_1, c_2) = 0$), but for nonhomogeneous (nonzero) ICs. The conclusion for stability (no blow-up) is the same as stated here, that is, the repellent term must equal or exceed the attraction term in eqs. (2.3a), (2.4a).

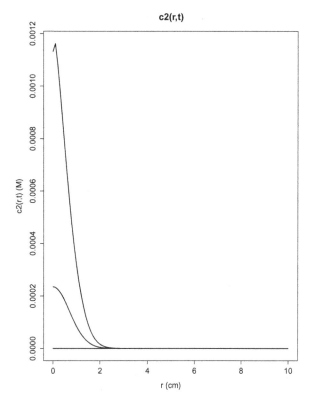

Figure 5.1c: $c_2(r,t)$ against r with t as a parameter, `chi2=0`

As a variant of Listing 5.1, 3D plotting of the solutions $m(r,t), c_1(r,t) c_2(r,t)$ can be added with the utility `persp`. The changes in Listing 5.1 are detailed next.

- $\chi_1 = \chi_2 = 1$ for balancing of attraction and repulsion is considered.

```
#
# Parameters
  mu=3.0e-09;d1=1.0e-07;d2=1.0e-07;
  chi1=1;chi2=1;
  a1=1.0e-12;a2=1.0e-12;
  b1=5.0e-04;b2=5.0e-04;
  r1=1.0e+04;r2=1;
```

- Calls to **persp** are added at the end of the main program for plotting $m(r, t), c_1(r, t), c_2(r, t)$.

```
#
# 3D
#
# m
    par(mfrow=c(1,1));
    persp(r,tout,m,theta=45,phi=45,xlab="r",
        ylab="t",zlab="m(r,t)",main="m(r,t)");
#
# c1
    par(mfrow=c(1,1));
    persp(r,tout,c1,theta=45,phi=45,xlab="r",
        ylab="t",zlab="c1(r,t)",main="c1(r,t)");
#
# c2
    par(mfrow=c(1,1));
    persp(r,tout,c2,theta=45,phi=45,xlab="r",
        ylab="t",zlab="c2(r,t)",main="c2(r,t)");
```

Note that the two independent variables in the base $(x - y)$ plane are r and t. $m(r, t)$ is the dependent variable (along the z axis) in the first plot. Similarly, $c_1(r, t), c_2(r, t)$ are used in the second and third plots. The resulting output is in Figs. 5.2a, 5.2b and 5.2c.

The 3D plots generally conform to the 2D plots of Figs. 4.2a, 4.2b and 4.2c (monotonically increasing solutions in response to the source term **f1** in Listing 3.1).

To summarize this section, MOL analysis is not a mechanical procedure with a guaranteed successful outcome. Rather, a new problem may require some experimentation to achieve a final stable numerical solution with acceptable accuracy. Specifically, h and p refinement (in both space and time) can be

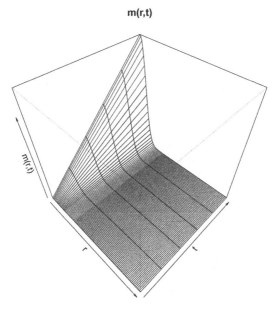

Figure 5.2a: $m(r, t)$ against r, t, `chi1=1`; `chi2=1`

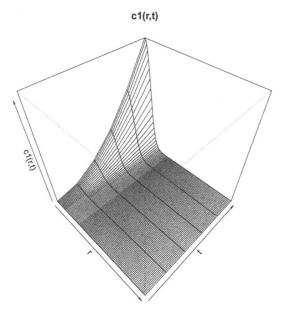

Figure 5.2b: $c_1(r, t)$ against r, t, `chi1=1`; `chi2=1`

c2(r,t)

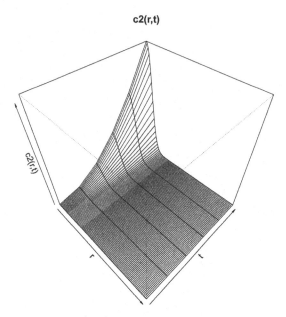

Figure 5.2c: $c_2(r, t)$ against r, t, `chi1=1`; `chi2=1`

used to establish apparent accuracy of the numerical solutions. The p refinement might include the programming of alternative ODE integration procedures, e.g, Euler and `lsodes` integrators.

(5.2) Alternate Boundary Conditions

Homogeneous, Neumann BCs (2.5) have been used throughout the preceding solutions. *Dirichlet* BCs would be another possibility which would anchor the solution at the boundary $r = r_0$ to a prescribed value, but this does not seem realistic for solutions with increasing values of $m(r, t), c_1(r, t), c_2(r, t)$. *Robin* (third-type) BCs could also be considered which consist of a linear combination of Dirichlet and Neumann BCs. Specifically,

$$\mu \frac{\partial m(r = r_0, t)}{\partial r} = k_m(m_b - m(r = r_0, t)) \qquad (5.1a)$$

$$D_1 \frac{\partial c_1(r = r_0, t)}{\partial r} = k_1(c_{1b} - c_1(r = r_0, t)) \tag{5.1b}$$

$$D_2 \frac{\partial c_2(r = r_0, t)}{\partial r} = k_2(c_{2b} - c_2(r = r_0, t)) \tag{5.1c}$$

where m_b, c_{1b}, c_{2b} are boundary or ambient values of $m(r,t)$, $c_1(r,t)$, $c_2(r,t)$ and k_m, k_1, k_2 can be considered as mass transfer coefficients.

BCs 5.1 are implemented in the following ODE/MOL routine which is a minor variant of the routine in Listing 3.1.

```
  pde_1c=function(t,u,parms){
#
# Function pde_1c computes the t derivative
# vectors of m(r,t),c1(r,t),c2(r,t)
#
# One vector to three vectors
  m=rep(0,nr);c1=rep(0,nr);c2=rep(0,nr);
  for(i in 1:nr){
     m[i]=u[i];
     c1[i]=u[i+nr];
     c2[i]=u[i+2*nr];
  }
#
# mr,c1r,c2r
   mr=dss006(0,r0,nr, m);
   c1r=dss006(0,r0,nr,c1);
   c2r=dss006(0,r0,nr,c2);
#
# BCs
   mr[1]=0;   mr[nr]=(km/mu)*( mb -m[nr]);
   c1r[1]=0;  c1r[nr]=(k1/d1)*(c1b-c1[nr]);
   c2r[1]=0;  c2r[nr]=(k2/d2)*(c2b-c2[nr]);
```

```
#
# mrr,c1rr,c2rr
  mrr=dss006(0,r0,nr, mr);
  c1rr=dss006(0,r0,nr,c1r);
  c2rr=dss006(0,r0,nr,c2r);
#
# PDEs
  mt=rep(0,nr);c1t=rep(0,nr);c2t=rep(0,nr);
  for(i in 1:nr){
    if(i==1){
      mt[i]=3*mu*mrr[i]-chi1*3*m[i]*c1rr[i]+
                        chi2*3*m[i]*c2rr[i]+
                        f1(r[i]);
      c1t[i]=3*d1*c1rr[i]+a1*m[i]-b1*c1[i];
      c2t[i]=3*d2*c2rr[i]+a2*m[i]-b2*c2[i];
    }
    if(i>1){
      mt[i]=mu*(mrr[i]+(2/r[i])*mr[i])-
            chi1*(m[i]*c1rr[i]+mr[i]*c1r[i]+
            (2/r[i])*m[i]*c1r[i])+
            chi2*(m[i]*c2rr[i]+mr[i]*c2r[i]+
            (2/r[i])*m[i]*c2r[i])+
            f1(r[i]);
      c1t[i]=d1*(c1rr[i]+(2/r[i])*c1r[i])+
             a1*m[i]-b1*c1[i];
      c2t[i]=d2*(c2rr[i]+(2/r[i])*c2r[i])+
             a2*m[i]-b2*c2[i];
    }
  }
#
# Three vectors to one vector
  ut=rep(0,3*nr);
  for(i in 1:nr){
    ut[i]        =mt[i];
```

```
   ut[i+nr]   =c1t[i];
   ut[i+2*nr]=c2t[i];
  }
#
# Increment calls to pde_1c
  ncall <<- ncall+1;
#
# Return derivative vector
  return(list(c(ut)));
  }
```

Listing 5.2: ODE/MOL routine `pde_1c` for Robin BCs

We can note the following details about Listing 5.2.

- `pde_1c` is similar to `pde_1a` of Listing 3.1. In particular, it is called by `lsodes` in the main program discussed subsequently.
- The essential difference in `pde_1c` is the use of the Robin BCs of eqs. (5.1) (in place of the homogeneous Neumann BCs of eqs. (2.5)).

```
#
# BCs
   mr[1]=0;   mr[nr]=(km/mu)*( mb -m[nr]);
  c1r[1]=0;  c1r[nr]=(k1/d1)*(c1b-c1[nr]);
  c2r[1]=0;  c2r[nr]=(k2/d2)*(c2b-c2[nr]);
```

The symmetry BCs at $r = 0$ are retained (using the subscript 1). The Robin BCs (5.1) are implemented at the boundary point $r = r_0$ (using the subscript `nr`). These BCs require the definition of `km,k1,k2` and `mb,c1b,c2b` in the main program.

- The derivative vector is returned as a list as required by `lsodes`.

```
#
# Return derivative vector
  return(list(c(ut)));
```

The main program is similar to the one in Listing 3.2. The changes in Listing 3.2 are detailed next.

- ODE/MOL routine of Listing 5.2 is accessed.

```
#
# Access functions for numerical solution
  setwd("f:/neuro/chap5");
  source("pde_1c.R");source("f1.R");
  source("dss006.R");
  source("dss008.R");
```

- The parameters `km,k1,k2` and `mb,c1b,c2b` are defined numerically (and passed to `pde_1c` of Listing 5.2).

```
#
# Parameters
  mu=3.0e-09;d1=1.0e-07;d2=1.0e-07;
  km=mu;k1=d1;k2=d2;
  mb=0;c1b=0;c2b=0;
  chi1=1;chi2=1;
  a1=1.0e-12;a2=1.2e-12;
  b1=5.0e-04;b2=5.0e-04;
  r1=1.0e+04;r2=1;
```

For these parameters, the cells and chemicals move into the surrounding region ($r > r_0$) that has no cells or chemicals (mb=c1b=c2b=0). The mass transfer coefficients are equated to the diffusivities since alternative values are not available (km=mu;k1=d1;k2=d2;).

- pde_1c is called by `lsodes`.

```
#
# ODE integration
  out=lsodes(y=u0,times=tout,func=pde_1c,
      sparsetype ="sparseint",rtol=1e-6,
      atol=1e-6,maxord=5);
  nrow(out)
  ncol(out)
```

The remainder of the main program is the same as in Listing 3.2.

The numerical and graphical output is esentially the same as in Table 4.1b and Figs. 4.2a, 4.2b, 4.2b and therefore is not presented here to conserve space (this is not unexpected since the solutions at $r = r_0$ remain at the zero initial value, i.e., the BCs at $r = r_0$ have no effect). The discussion of BCs (5.1) was included to demonstrate how alternate BCs (e.g., rather than homogeneous Neumann BCs (2.5)) can be implemented within the MOL framework. The variations in BCs might be useful in continuing studies of the chemotaxis model, particularly as the solutions approach the boundary at $r = r_0$.

(5.3) Time-limited Source Function

As a final variant of the chemotaxis model of eqs. (2.3) to (2.6), a cell volumetric source term in eqs. (2.3a), (2.4a) that varies with t is considered. As a first example, a step in t is programmed in function f2, a variant of the earlier f1 in Listing 3.3.

```
  f2=function(r,t){
#
# Function f2 computes the inhomogeneous
# volumetric source term of the m(r,t) PDE
#
# No source of cells
# f2=0;
#
```

```
# Source for a limited time
  if(t<=t1){f2=r1*exp(-r2*r^2);}
  if(t> t1){f2=0;}
#
# Source from a pulse
# if(t < t1)                {f2=0;}
# if((t>=t1)&(t<=(t1+t2))){f2=r1*exp(-r2*r^2);}
# if(t>(t1+t2))             {f2=0;}
#
# Return f2
  return(c(f2));
  }
```

Listing 5.3: Cell source as a function of t

We can note the following details about Listing 5.3.

- A second input argument, t, has been added to f1 of Listing 3.3.

  ```
    f2=function(r,t){
  #
  # Function f2 computes the inhomogeneous
  # volumetric source term of the m(r,t) PDE
  ```

- The case of a zero source term can be selected by deactivating the third comment in

  ```
  #
  # No source of cells
  # f2=0;
  ```

  Execution of this case gives the time-invariant solution discussed in Chapter 4 (dependent variables $m(r,t), c_1(r,t)$, $c_2(r,t)$ remain at the zero IC).

- A step in the source term at `t=t1` is executed with

```
#
# Source for a limited time
  if(t<=t1){f2=r1*exp(-r2*r^2);}
  if(t> t1){f2=0;}
```

t is available as the second argument of `f2` and `t1` is set in the main program (discussed next).
- The source term as a square pulse in *t* can be selected by deactivating the comments in the executable statements in

```
#
# Source from a pulse
# if(t < t1)                    {f2=0;}
# if((t>=t1)&(t<=(t1+t2))){f2=r1*exp(-r2*r^2);}
# if(t>(t1+t2))                 {f2=0;}
```

The duration of the pulse, `t2`, is set in the main program.
- The source term is returned to the ODE/MOL routine `pde_1d`[5] called by `lsodes` in the main program.

```
#
# Return f2
  return(c(f2));
  }
```

The main program is from Listing 3.2 with the following modifications.

- Files `pde_1d,  f2` are accessed for the particular case of a time-varying source term

```
#
# Access functions for numerical solution
  setwd("f:/neuro/chap5");
```

---

[5]`f2(r[i],t)` is called in place of `f1(r[i])` in `pde_1d`.

```
source("pde_1d.R");source("f2.R");
source("dss006.R");
source("dss008.R");
```

- The parameters are defined numerically with

```
#
# Parameters
  mu=3.0e-09;d1=1.0e-07;d2=1.0e-07;
  chi1=1;chi2=1;
  a1=1.0e-12;a2=1.2e-12;
  b1=5.0e-04;b2=5.0e-04;
  r1=1.0e+04;r2=1;
  t1=500;t2=100;
```

Note in particular the values of t1,t2 that are then used in f2.

- The interval in t is extended to $0 \le t \le 3000$ to give a more complete transient (approach to an equilibrium or steady state) with nout=11 output points corresponding to output at $t = 0, 300, \dots, 3000$ sec or $t = 0, 5, \dots, 50$ min.

```
#
# Independent variable for ODE integration
  t0=0;tf=3.0e+03;nout=11;
  tout=seq(from=t0,to=tf,by=(tf-t0)/(nout-1));
```

- pde_1d is called in an Euler integration (Listing 3.2 is based on lsodes).

```
#
# ODE integration
    m=matrix(0,nrow=nr,ncol=nout);
  c1=matrix(0,nrow=nr,ncol=nout);
  c2=matrix(0,nrow=nr,ncol=nout);
  u=u0;t=t0;h=2;
  for(it in 1:nout){
```

```
    for(ir in 1:nr){
       m[ir,it]=u[ir];
       c1[ir,it]=u[ir+nr];
       c2[ir,it]=u[ir+2*nr];
    }
    if(it==nout)break;
    for(i in 1:100){
      derv=pde_1d(t,u,parm);
      u=u+derv*h;
      t=t+h;
    }
  }
```

- 3D plotting is added at the end

```
#
# 3D
#
# m
    par(mfrow=c(1,1));
    persp(r,tout,m,theta=45,phi=45,xlab="r",
          ylab="t",zlab="m(r,t)",main="m(r,t)");
#
# c1
    par(mfrow=c(1,1));
    persp(r,tout,c1,theta=45,phi=45,xlab="r",
          ylab="t",zlab="c1(r,t)",main="c1(r,t)");
#
# c2
    par(mfrow=c(1,1));
    persp(r,tout,c2,theta=45,phi=45,xlab="r",
          ylab="t",zlab="c2(r,t)",main="c2(r,t)");
```

Abbreviated numerical output (for a step change in the source term in f2) follows.

t	r	m(r,t)	c1(r,t)	c2(r,t)
0.00	0.0	0.000e+00	0.000e+00	0.000e+00
0.00	0.1	0.000e+00	0.000e+00	0.000e+00
0.00	0.2	0.000e+00	0.000e+00	0.000e+00
0.00	0.3	0.000e+00	0.000e+00	0.000e+00
0.00	0.4	0.000e+00	0.000e+00	0.000e+00
0.00	0.5	0.000e+00	0.000e+00	0.000e+00
.				.
.				.
.				.

Output for r = 0.6 to 9.4 removed

.				.
.				.
.				.
0.00	9.5	0.000e+00	0.000e+00	0.000e+00
0.00	9.6	0.000e+00	0.000e+00	0.000e+00
0.00	9.7	0.000e+00	0.000e+00	0.000e+00
0.00	9.8	0.000e+00	0.000e+00	0.000e+00
0.00	9.9	0.000e+00	0.000e+00	0.000e+00
0.00	10.0	0.000e+00	0.000e+00	0.000e+00

t	r	m(r,t)	c1(r,t)	c2(r,t)
50.00	0.0	7.174e+05	2.249e-03	2.699e-03
50.00	0.1	7.188e+05	2.243e-03	2.691e-03
50.00	0.2	7.230e+05	2.224e-03	2.669e-03
50.00	0.3	7.297e+05	2.193e-03	2.632e-03
50.00	0.4	7.390e+05	2.150e-03	2.580e-03
50.00	0.5	7.504e+05	2.096e-03	2.515e-03
.				.
.				.
.				.

Output for r = 0.6 to 9.4 removed

.				.
.				.

	·			·
50.00	9.5	3.217e-33	3.972e-42	4.767e-42
50.00	9.6	4.766e-34	5.910e-43	7.092e-43
50.00	9.7	6.926e-35	8.733e-44	1.048e-43
50.00	9.8	9.683e-36	9.417e-45	1.130e-44
50.00	9.9	1.379e-36	1.874e-45	2.249e-45
50.00	10.0	-2.641e-37	-8.015e-45	-9.618e-45

```
ncall =   1000
```

Table 5.1: Abbreviated output for a step change in the source term $f(m, c_1, c_2)$

ncall = (100)(11-1) = 1000, with the final t = 50.00, as expected.

The graphical output is in Figs. 5.3a, 5.3b, 5.3c. The effect of the step in the source term in f2 is clear. Also, the nonzero

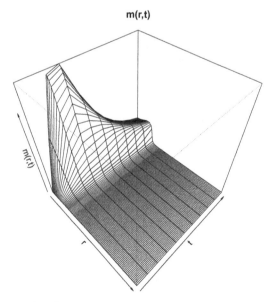

m(r,t)

Figure 5.3a: $m(r, t)$ against r, t, chi1=1; chi2=1, step source in f2

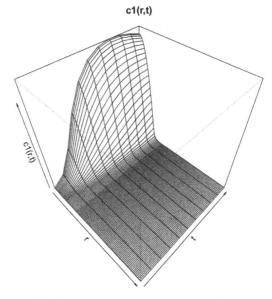

Figure 5.3b: $c_1(r,t)$ against r, t, `chi1=1`; `chi2=1`, step source in `f2`

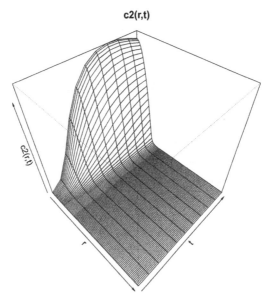

Figure 5.3c: $c_2(r,t)$ against r, t, `chi1=1`; `chi2=1`, step source in `f2`

portion of the solution is largely confined to $r < 2$ as noted previously. An increase in the Fickian diffusion (increase in μ, D_1, D_2 in eqs. (2.3), (2.4)) should give a spreading and smoothing of the solutions $m(r, t), c_1(r, t), c_2(r, t)$.

(5.4) Summary and Conclusions

The preceding discussions demonstrate possible approaches to experimenting with the chemotaxis model. Clearly many possibilities can be considered and the reader can explore these and others that come to mind.

As a word of caution, when the model is changed, (e.g., variations in the parameters, ICs and BCs, terms in the PDEs, addition of PDEs) the execution of the routines with reasonable accuracy and computer times cannot be guaranteed in advance. Systems of PDEs can be unexpectedly sensitive to changes, particularly in the presence of nonlinearties such as the attraction and repulsion chemotaxis terms.

Thus, some trial-and-error, e.g., with the space and time scales (in r and t), the number of spatial points, the ODE integrator error tolerances, may be required to achieve satisfactory execution of the model. Displaying the t derivatives and PDE RHS terms computed in the ODE/MOL routine can provide insights into why numerical problems are encountered, such as terms that are unrealistically large or of the wrong sign. The judgement of the analyst with respect to knowledge of the model and computer code is essential for this type of investigation and verification.

This concludes the discussion of the output for a series of variants of the model. The intent is to demonstrate how the model can be changed and executed to study various effects (e.g., parameter changes, instability or blow-up, effect of changes in the cell source term).

Extensions of the model can be considered and implemented using the MOL methodology discussed previously. For example, the Fickian diffusivities could be considered as a function of the PDE dependent variables, i.e., $\mu(m, c_1, c_2)$, $D_1(m, c_1, c_2)$, $D_2(m, c_1, c_2)$. Also, the chemotactic diffusivities could be considered as functions of the dependent variables, i.e., $\chi_1(m, c_1, c_2)$, $\chi_2(m, c_1, c_2)$. The chemotactic diffusivities could also be considered as a function of the dependent variable gradients,

$$\chi_1\left(\frac{\partial m}{\partial r}, \frac{\partial c_1}{\partial r}, \frac{\partial c_2}{\partial r}\right), \quad \chi_2\left(\frac{\partial m}{\partial r}, \frac{\partial c_1}{\partial r}, \frac{\partial c_2}{\partial r}\right)$$

These possibilities are left for consideration, and possible implementation (coding), by the reader. Additional extensions are mentioned in the concluding chapter that follows.

Reference

[1]   Lin, K., and C. Mu (2016), Global existence and convergence to steady states for an attraction-repulsion chemotaxis system, *Nonlinear Analysis: Real World Applications*, **31**, 630–643

Chapter 6

Extensions and Conclusions

We conclude the discussion of chemotaxis model development with some extensions of the preceding models.

(6.1) Alternate Production/Depletion Rates

As noted in Chapter 2, with

$$g_1(m, c_1, c_2) = a_1 m - b_1 c_1, \; g_2(m, c_1, c_2) = a_2 m - b_2 c_2$$

functions g_1, g_2 represent first-order (linear) reaction terms based on m (a forward or production reaction) and c_1, c_2 (for reverse or consumption reactions) with constants a_1, a_2, b_1, b_2. These functions are specific selections as suggested in [2], but they can be of essentially any form. For example, the nonlinear forms

$$g_1(m, c_1, c_2) = (a_1 - b_1)mc_1, \; g_2(m, c_1, c_2) = (a_2 - b_2)mc_2$$

can be considered where $a_1 - b_1 > 0, a_2 - b_2 > 0$ corresponds to production of c_1, c_2 while $a_1 - b_1 < 0, a_2 - b_2 < 0$ corresponds to depletion or consumption. Whether this form of g_1, g_2 is applicable will depend on the judgement of the analyst as the model is developed, but they can be easily programmed in the ODE/MOL routine. Thus, experimentation with alternate forms of the model is readily accomplished numerically, including nonlinear forms.

(6.2) Nonlinear Diffusion

If the diffusivities μ, D_1, D_2 are functions of the PDE dependent variables, i.e., $\mu = \mu(m, c_1, c_2)$, $D_1 = D_1(m, c_1, c_2)$, $D_2 = D_2(m, c_1, c_2)$, the Fickian diffusion terms in eqs. (2.3) and (2.4) become

$$\frac{1}{r^2}\frac{\partial\left(\mu r^2 \frac{\partial m}{\partial r}\right)}{\partial r}, \quad \frac{1}{r^2}\frac{\partial\left(D_1 r^2 \frac{\partial c_1}{\partial r}\right)}{\partial r}, \quad \frac{1}{r^2}\frac{\partial\left(D_2 r^2 \frac{\partial c_2}{\partial r}\right)}{\partial r}$$

These terms can be included within the numerical solution. For example, for the $\mu = \mu(m, c_1, c_2)$ term,

$$\frac{1}{r^2}\frac{\partial\left(\mu r^2 \frac{\partial m}{\partial r}\right)}{\partial r}$$

$$= \mu\frac{\partial^2 m}{\partial r^2} + \frac{2}{r}\mu\frac{\partial m}{\partial r} + \left(\frac{\partial\mu}{\partial m}\frac{\partial m}{\partial r} + \frac{\partial\mu}{\partial c_1}\frac{\partial c_1}{\partial r} + \frac{\partial\mu}{\partial c_2}\frac{\partial c_2}{\partial r}\right)\frac{\partial m}{\partial r}$$

For $\mu = const$, $\dfrac{\partial\mu}{\partial m} = \dfrac{\partial\mu}{\partial c_1} = \dfrac{\partial\mu}{\partial c_2} = 0$ and the result is the form used previously in Chapter 2.

$$\frac{1}{r^2}\frac{\partial\left(\mu r^2 \frac{\partial m}{\partial r}\right)}{\partial r} = \mu\frac{\partial^2 m}{\partial r^2} + \frac{2}{r}\mu\frac{\partial m}{\partial r}$$

As an example of a variable diffusivity, if $\mu = 1 + m$, the Fickian diffusion term in eqs. (2.3a) and (2.4a) is

$$\frac{1}{r^2}\frac{\partial\left(\mu r^2 \frac{\partial m}{\partial r}\right)}{\partial r} = (1+m)\frac{\partial^2 m}{\partial r^2} + \frac{2}{r}(1+m)\frac{\partial m}{\partial r} + (1+0+0)\frac{\partial m}{\partial r}$$

$$= (1+m)\frac{\partial^2 m}{\partial r^2} + \left(\frac{2}{r}(1+m) + 1\right)\frac{\partial m}{\partial r}$$

The nonlinearity of this last result is clear, but it can easily be programmed in the ODE/MOL routine.

Similar considerations apply to the chemotactic diffusion terms for $\chi_1 = \chi_1(m, c_1, c_2)$, $\chi_2 = \chi_2(m, c_1, c_2)$. Thus, the use of variable diffusivities is straightforward (but the successful execution of the resulting code cannot be guaranteed in advance).

(6.3) Additional Cells and Chemicals

Additional cells and chemicals can be added to the model. For example, for two types of cells and four chemicals, the code at the beginning of the ODE/MOL routine would be

```
  pde_1a=function(t,u,parms){
#
# Function pde_1a computes the t derivative
# vectors of m1(r,t),m2(r,t),c1(r,t),c2(r,t),
# c3(r,t),c4(r,t)
#
# One vector to six vectors
  m1=rep(0,nr);m2=rep(0,nr);
  c1=rep(0,nr);c2=rep(0,nr);
  c3=rep(0,nr);c4=rep(0,nr);
  for(i in 1:nr){
    m1[i]=u[i];m2[i]=u[i+nr];
    c1[i]=u[i+2*nr];c2[i]=u[i+3*nr];
    c3[i]=u[i+4*nr];c4[i]=u[i+5*nr];
  }
```

and at the end of the ODE/MOL routine (after calculation of the derivatives in t)

```
#
# Six vectors to one vector
  ut=rep(0,6*nr);
  for(i in 1:nr){
    ut[i]      =m1t[i];ut[i+nr]   =m2t[i];
```

```
    ut[i+2*nr]=c1t[i];ut[i+3*nr]=c2t[i];
    ut[i+4*nr]=c3t[i];ut[i+5*nr]=c4t[i];
  }
```

(6.4) Conclusions

The preceding examples demonstrate that the numerical MOL approach is flexible and extensible. The resulting code(s) can be used for computer-based experimentation to observe the properties of the solutions and to eventually reach a formulation that provides insight into the PDE model characteristics, and ideally, an interpretation of available experimental data and suggestions for ND (e.g., AD)[1] therapeutic treatment. This would probably start with an association of the cells, e.g., microglia cells, and chemicals (proteins), e.g., β-amyloids[2] and Interleuken 1 and 6, as attractants and repellents in the model. Variations in the model would be directed by existing knowledge of ND and serve as a guide for continuing experimental and epidemiological studies.

The limitations of the model include:

- Spatial-averaging (a continuum model) of the cell population and chemical concentrations. That is, fine spatial structure is not attempted and spatial symmetry around $r = 0$ is assumed.
- Only a single source of cells (centered at $r = 0$) is considered.
- Long-term ND effects (e.g., of AD) are not considered. Rather, short term conditions that might lead to the observable effects of ND are the main focus.

In appendix A1 that follows, additional consideration is given to: (1) the accuracy of the numerical approximation of the chemotaxis terms in eqs. (2.3a), (2.4a), and (2) the effect of these terms on the solution of eqs. (2.3) and (2.4).

---

[1]ND $\Rightarrow$ neurodegenerative disease; AD $\Rightarrow$ Alzheimer's disease.

[2]Additional information about β-amyloid kinetics is available in [3].

References

[1] Lin, K., and C. Mu (2016), Global existence and convergence to steady states for an attraction-repulsion chemotaxis system, *Nonlinear Analysis: Real World Applications*, **31**, 630–643

[2] Luca, M., A. Chavez-Ross, L. Edelstein-Keshet and A. Mogilner (2003), Chemotactic signaling, microglia, and Alzheimer's disease senile plaques: Is there a connection?, *Bulletin of Mathematical Biology*, **65**, 693–730

[3] Morelli, S., et al (2016), Neuronal membrane bioreactor as a tool for testing crocin neuroprotective effect in Alzheimer's disease, *Chemical Engineering Journal*, **305**, 69–78

Appendix A: Analysis of PDE Chemotaxis Terms

The nonlinear chemotaxis terms of eqs. (2.3), (2.4) are a distinctive feature of the PDE model considered in the preceding chapters. Therefore, the numerical approximation and properties of these terms are of particular interest when considering the model. Here we discuss these two topics in terms of a numerical analysis.

(A.1) Numerical Approximation of PDE Chemotaxis Terms

The chemotaxis terms considered next are

Attractant $c_1(x, t)$:

$$q_a = \chi_1 \left(m \frac{\partial c_1}{\partial x} \right) \tag{A.1a}$$

Repellent $c_2(x, t)$:

$$q_r = -\chi_2 \left(m \frac{\partial c_2}{\partial x} \right) \tag{A.1b}$$

where q_a, q_r are the fluxes from the attractant and repellent, respectively. Eqs. (A.1) are 1D in the Cartesian coordinate x to simplify the analysis (rather than the terms in r of eqs. (2.3a),

(2.4a) with the variable coefficient in r and the singularity $1/r$ for $r = 0$).

To reiterate (from Chapters 1,2), these chemotaxis diffusion flux terms are nonlinear, e.g., from the product $m\dfrac{\partial c_1}{\partial x}$ with two components m, c_1 in the q_a flux of eq. (A.1a). Therefore, the complexity of the PDE system precludes an analytical solution that can be used to verify the numerical solution.

An alternate approach to verifying the computed chemotaxis terms is to assume a x variation of the three dependent variables $m(x, t), c_1(x, t), c_2(x, t)$, then compare the numerical values of the chemotaxis terms with the analytical values. This is not a proof that the numerical terms are correct, but rather, indicates the level of agreement between the numerical and analytical values for the assumed x variation.

To illustrate this approach, the following routine is based on an assumed exponential variation of the three dependent variables.

```
#
# Previous workspaces are cleared
  rm(list=ls(all=TRUE))
#
# Access files
  setwd("f:/neuro/app1");
  source("dss006.R");
  source("dss008.R");
#
# Define uniform grid
  xl=0;xu=1;n=21;
  x=seq(from=xl,to=xu,by=(xu-xl)/(n-1));
#
# Define dependent variables
  a=-1;b=-1;c=-1;
```

```
  chi1=1;chi2=1;
   m=exp(a*x);
  c1=exp(b*x);
  c2=exp(c*x);
#
# c1x,c2x gradients
  c1x=dss006(xl,xu,n,c1);
  c2x=dss006(xl,xu,n,c2);
#
# Approximation of chemotaxis fluxes
  qa= chi1*m*c1x;
  qr=-chi2*m*c2x;
  netx=qa+qr;
#
# Exact fluxes
  qae= chi1*exp(a*x)*b*exp(b*x);
  qre=-chi2*exp(a*x)*c*exp(c*x);
  netxe=qae+qre;
#
# Display comparison of exact fluxes and approximations
  cat(sprintf("\n     x         qa          qr
              netx "));
  cat(sprintf("\n     x         qae         qre
              netxe"));
  iv=seq(from=1,to=n,by=5);
  for(i in iv){
    cat(sprintf("\n%5.3f%10.5f%10.5f%12.7f",
        x[i], qa[i], qr[i], netx[i]));
    cat(sprintf("\n%5.3f%10.5f%10.5f%12.7f\n",
        x[i],qae[i],qre[i],netxe[i]));
  }
#
# Approximation of second derivatives
  qax=-dss006(xl,xu,n,qa);
```

```
  qrx=-dss006(xl,xu,n,qr);
  netxx=qax+qrx;
#
# Exact second derivatives
  qaxe=-chi1*(exp(a*x)*b^2*exp(b*x)+
          b*exp(b*x)*a*exp(a*x));
  qrxe= chi2*(exp(a*x)*c^2*exp(c*x)+
          c*exp(c*x)*a*exp(a*x));
  netxxe=qaxe+qrxe;
#
# Display comparison of second derivatives
# and approximations
  cat(sprintf("\n    x       qax       qrx
              netxx "));
  cat(sprintf("\n    x       qaxe      qrxe
              netxxe"));
  iv=seq(from=1,to=n,by=5);
  for(i in iv){
    cat(sprintf("\n%5.3f%10.5f%10.5f%12.7f",
        x[i], qax[i], qrx[i], netxx[i]));
    cat(sprintf("\n%5.3f%10.5f%10.5f%12.7f\n",
        x[i],qaxe[i],qrxe[i],netxxe[i]));
  }
#
# Plot fluxes, second derivatives
#
# qa
  matplot(x,qa,type="l",lwd=2,col="black",lty=1,
    xlab="x",ylab="qa",main="qa");
  matpoints(x,qae,pch="o",col="black");
#
# qr
  matplot(x,qr,type="l",lwd=2,col="black",lty=1,
    xlab="x",ylab="gr",main="qr");
```

```
  matpoints(x,qre,pch="o",col="black");
#
# netx
  matplot(x,netx,type="l",lwd=2,col="black",lty=1,
    xlab="x",ylab="netx",main="netx");
  matpoints(x,netx,pch="o",col="black");
#
# qax
  matplot(x,qax,type="l",lwd=2,col="black",lty=1,
    xlab="x",ylab="-(qa)x",main="-(qa)x");
  matpoints(x,qaxe,pch="o",col="black");
#
# qrx
  matplot(x,qrx,type="l",lwd=2,col="black",lty=1,
    xlab="x",ylab="-(qr)x",main="-(qr)x");
  matpoints(x,qrxe,pch="o",col="black");
#
# netxx
  matplot(x,netxx,type="l",lwd=2,col="black",lty=1,
    xlab="x",ylab="netxx",main="netxx");
  matpoints(x,netxx,pch="o",col="black");
```

Listing A.1: Main program to elucidate the chemotaxis fluxes

We can note the following details about Listing A.1.

- Previous workspaces (files) are cleared and routines used below are accessed.

```
  #
  # Previous workspaces are cleared
    rm(list=ls(all=TRUE))
  #
  # Access files
    setwd("f:/neuro/app1");
```

```
source("dss006.R");
source("dss008.R");
```

The `setwd` (set working directory) requires editing for the local computer. `dss006`, a spatial differentiation routine, is listed in Appendix B.

- A grid in x is defined with 21 points for $0 \le x \le 1$. Therefore, $x = 0, 0.05, \ldots, 1$.

```
#
# Define uniform grid
  xl=0;xu=1;n=21;
  x=seq(from=xl,to=xu,by=(xu-xl)/(n-1));
```

- The PDE dependent variables $m(x,t), c_1(x,t), c_2(x,t)$ (eqs. (2.3a), (2.4a)) are defined as exponential functions in x.

```
#
# Define dependent variables
  a=-1;b=-1;c=-1;
  chi1=1;chi2=1;
   m=exp(a*x);
  c1=exp(b*x);
  c2=exp(c*x);
```

These are vectorized equations since x is a 21-vector. The functions decay with x since $a < 0, b < 0, c < 0$ as expected since the volumentric source term `f1` is centered at the origin ($r = 0$ in eqs. (2.3a),(2.4a), $x = 0$ in eqs. (A.1)). Also, χ_1, χ_2 in eqs. (A.1) are set to one (so the attraction and repulsion are balanced).

- The gradients (derivatives, slopes) in eqs. (A.1) are computed with the library differentiator `dss006`.

Gradient in eqs. (A.1)   Code

$$\frac{\partial c_1}{\partial x}$$ `c1x=dss006(xl,xu,n,c1)`

$$\frac{\partial c_2}{\partial x}$$ `c2x=dss006(xl,xu,n,c2)`

That is,

```
#
# c1x,c2x gradients
  c1x=dss006(xl,xu,n,c1);
  c2x=dss006(xl,xu,n,c2);
```

• The fluxes of eqs. (A.1) are computed.

Eqs. (A.1)   Code

$$q_a = \chi_1\left(m\frac{\partial c_1}{\partial x}\right)$$ `qa=chi1*m*c1x`

$$q_r = -\chi_2\left(m\frac{\partial c_2}{\partial x}\right)$$ `qr=-chi2*m*c2x`

That is,

```
#
# Approximation of chemotaxis fluxes
  qa= chi1*m*c1x;
  qr=-chi2*m*c2x;
  netx=qa+qr;
```

Also, the net flux is computed, `netx=qa+qr`

- The exact fluxes are computed from the exponential functions.

Exponential functions	Code
$q_a = \chi_1 b e^{ax} e^{bx}$	qae=chi1*exp(a*x)*b*exp(b*x)
$q_r = -\chi_2 c e^{ax} e^{cx}$	qre=-chi2*exp(a*x)*c*exp(c*x)

That is,

```
#
# Exact fluxes
  qae= chi1*exp(a*x)*b*exp(b*x);
  qre=-chi2*exp(a*x)*c*exp(c*x);
  netxe=qae+qre;
```

Also, the net exact flux is computed, netxe=qae+qre
- The approximate (numerical) and exact fluxes are displayed.

```
#
# Display comparison of exact fluxes and approximatic
  cat(sprintf("\n     x          qa          qr
              netx "));
  cat(sprintf("\n     x          qae         qre
              netxe"));
  iv=seq(from=1,to=n,by=5);
  for(i in iv){
    cat(sprintf("\n%5.3f%10.5f%10.5f%12.7f",
        x[i], qa[i], qr[i], netx[i]));
    cat(sprintf("\n%5.3f%10.5f%10.5f%12.7f\n",
        x[i],qae[i],qre[i],netxe[i]));
  }
```

- The RHS terms in a material balance are computed.

$$\frac{\partial m}{\partial t} = -\frac{\partial q_a}{\partial x} - \frac{\partial q_r}{\partial x}$$

$$= -\chi_1 \frac{\partial}{\partial x}\left(m\frac{\partial c_1}{\partial x}\right) - \chi_2\frac{\partial}{\partial x}\left(-m\frac{\partial c_2}{\partial x}\right) \qquad \text{(A.2a)}$$

```
#
# Approximation of second derivatives
  qax=-dss006(xl,xu,n,qa);
  qrx=-dss006(xl,xu,n,qr);
  netxx=qax+qrx;
```

This material balance could contain other terms such as a Fickian diffusion term and a volumetric source term (as in eqs. (2.3a), (2.4a)), but here we are considering only the chemotaxis terms.

We can infer that the RHS terms in eq. (A.2) give stability or instability. For example, the diffusion equation

$$\frac{\partial u}{\partial t} = \pm D\frac{\partial^2 u}{\partial x^2} \qquad \text{(A.2b)}$$

is unstable for $D > 0$ and the minus in the RHS is used. Conversely, eq. (A.2b) is stable if the plus is used.

By analogy, the PDE with just the attraction term

$$\frac{\partial m}{\partial t} = -\chi_1\frac{\partial}{\partial x}\left(m\frac{\partial c_1}{\partial x}\right) \qquad \text{(A.2c)}$$

may be unstable (from the minus) and the PDE with just the repulsion term

$$\frac{\partial m}{\partial t} = \chi_2\frac{\partial}{\partial x}\left(m\frac{\partial c_2}{\partial x}\right) \qquad \text{(A.2d)}$$

may be stable.

This could possibly explain the instability discussed in Chapter 5 for chi1=1, chi2=0 (Figs. 5.1a,b,c), and the stability discussed in Chapter 4 for `chi1=1, chi2=1` (Figs. 4.2a,b,c). Physically, eq. (A.2c) could lead to an increased concentration of $m(x,t), c_1(x,t), c_2(x,t)$ (thereby leading to higher $c_1(x,t), c_2(x,t)$). If, for example, $c_1(x,t)$ represents β-amyloid, this increase could produce aggregation and then fibrils or plaque (and eventually AD).

Similarly, eq. (A.2d) could produce stable dispersion of $m(x,t), c_1(x,t), c_2(x,t)$ thereby leading to lower $c_1(x,t)$, $c_2(x,t)$. This balancing of attraction and repulsion would eventually lead to production of plaque (fibrils) or dispersion to prevent formation of plaque. This suggests as a therapy for AD the addition of a repulsive component so that the PDE system and physical system has a preponderance of repulsion.

This hypothesis can be studied further with the R routines that have been presented and discussed in Appendix A, and this is done briefly in the next section. However, we should keep in mind that this analysis is not rigorous since (1) it does not include t (it is static rather than dynamic), and (2) the full PDE model, eqs. (2.3), (2.4), is a coupled, nonlinear PDE system that can be studied numerically, as in Chapters 4,5, but would be difficult to study analytically.

- The exact second derivative terms in eq. (A.2a) are computed from the exponential functions.

```
#
# Exact second derivatives
  qaxe=-chi1*(exp(a*x)*b^2*exp(b*x)+
            b*exp(b*x)*a*exp(a*x));
  qrxe= chi2*(exp(a*x)*c^2*exp(c*x)+
            c*exp(c*x)*a*exp(a*x));
  netxxe=qaxe+qrxe;
```

Also, the net exact second derivative flux is computed, `netxxe=qaxe+qrxe`.

- The numerical and exact second derivative terms are displayed.

```
#
# Display comparison of second derivatives
# and approximations
  cat(sprintf("\n    x        qax         qrx
             netxx "));
  cat(sprintf("\n    x        qaxe        qrxe
             netxxe"));
  iv=seq(from=1,to=n,by=5);
  for(i in iv){
    cat(sprintf("\n%5.3f%10.5f%10.5f%12.7f",
        x[i], qax[i], qrx[i], netxx[i]));
    cat(sprintf("\n%5.3f%10.5f%10.5f%12.7f\n",
        x[i],qaxe[i],qrxe[i],netxxe[i]));
  }
```

- The fluxes are plotted.

```
#
# Plot fluxes, second derivatives
#
# qa
  matplot(x,qa,type="l",lwd=2,col="black",lty=1,
    xlab="x",ylab="qa",main="qa");
  matpoints(x,qae,pch="o",col="black");
#
# qr
  matplot(x,qr,type="l",lwd=2,col="black",lty=1,
    xlab="x",ylab="gr",main="qr");
  matpoints(x,qre,pch="o",col="black");
#
# netx
  matplot(x,netx,type="l",lwd=2,col="black",lty=1,
```

```
      xlab="x",ylab="netx",main="netx");
    matpoints(x,netx,pch="o",col="black");
```

The numerical fluxes are plotted as solid lines (using `matplot`) and the exact fluxes are plotted as points (using `matpoints`).

- The second derivative terms are plotted.

```
  #
  # qax
    matplot(x,qax,type="l",lwd=2,col="black",lty=1,
      xlab="x",ylab="-(qa)x",main="-(qa)x");
    matpoints(x,qaxe,pch="o",col="black");
  #
  # qrx
    matplot(x,qrx,type="l",lwd=2,col="black",lty=1,
      xlab="x",ylab="-(qr)x",main="-(qr)x");
    matpoints(x,qrxe,pch="o",col="black");
  #
  # netxx
    matplot(x,netxx,type="l",lwd=2,col="black",lty=1,
      xlab="x",ylab="netxx",main="netxx");
    matpoints(x,netxx,pch="o",col="black");
```

The numerical second derivatves are plotted as solid lines (using `matplot`) and the exact second derivatives are plotted as points (using `matpoints`).

Execution of the program of Listing A.1 gives the following output.

x	qa	qr	netx
x	qae	qre	netxe
0.000	-1.00000	1.00000	0.0000000
0.000	-1.00000	1.00000	0.0000000

0.250	-0.60653	0.60653	0.0000000
0.250	-0.60653	0.60653	0.0000000
0.500	-0.36788	0.36788	0.0000000
0.500	-0.36788	0.36788	0.0000000
0.750	-0.22313	0.22313	0.0000000
0.750	-0.22313	0.22313	0.0000000
1.000	-0.13534	0.13534	0.0000000
1.000	-0.13534	0.13534	0.0000000
x	qax	qrx	netxx
x	qaxe	qrxe	netxxe
0.000	-2.00000	2.00000	0.0000000
0.000	-2.00000	2.00000	0.0000000
0.250	-1.21306	1.21306	0.0000000
0.250	-1.21306	1.21306	0.0000000
0.500	-0.73576	0.73576	0.0000000
0.500	-0.73576	0.73576	0.0000000
0.750	-0.44626	0.44626	0.0000000
0.750	-0.44626	0.44626	0.0000000
1.000	-0.27067	0.27067	0.0000000
1.000	-0.27067	0.27067	0.0000000

Table A.1: Output from the program of Listing A.1

We can note the following details about this output.

- Every fifth value of the solution is displayed so that the interval in x is $5(0.05) = 0.25$.
- The numerical and exact fluxes and second derivatives agree to five figures. This indicates the calculation of the numerical chemotaxis terms in eqs. (2.3), (2.4) is sufficiently accurate to give numerical solutions to the PDE model of good accuracy. This results primarily from the approximations in dss006 which are seven point finite differences (FDs) (dss006 is listed in Appendix B).

    This accuracy is also due in part to the smooth exponential functions used to define $m(x,t), c_1(x,t), c_2(x,t)$. That is, functions in x that are less smooth can be expected to produce numerical chemotaxis terms with reduced accuracy.

    This accuracy can be further verified by using dss008 based on nine point FD approximations. In this case, the same numerical solutions (to five figures) are produced. This variation in the order of the FD approximations is termed p refinement in accordance with a truncation error of the form $O(\Delta x^p)$ where Δx is the grid spacing in x and $p = 6, 8$ for dss006, dss008, respectively.

- The attractive and repulsive chemotaxis terms are the same in absolute magnitude, but opposite in sign, so they sum to zero. The difference in the sign is due to the definition of the same exponential function for $m(x,t), c_1(x,t), c_2(x,t)$ and $\chi_1 = \chi_2 = 1$, but the sign change in eqs. (A.1).

    The graphical output of Figs. A.1a,b,c,d,e,f reflect the numerical output in Table A.1, but give a more complete indication of the variation of the fluxes and second derivative terms in x. As expected, the variations are greater near $x = 0$ due to the exponential functions and the dependence of the chemotaxis terms on the gradients (slopes) as reflected in eqs. (A.1). This

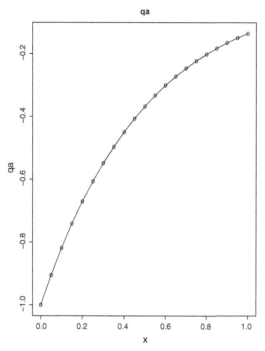

Figure A.1a: $q_a = \chi_1\left(m\dfrac{\partial c_1}{\partial x}\right)$ against x line - num, points - exact

is also the case for eqs. (2.3), (2.4) since the source term $f_1(r)$ is a Gaussian function centered at $r = 0$.

Again, the primary objective is using the program in List-ing A.1 is to confirm the apparent accuracy of the calculated chemotaxis terms in eqs. (2.3), (2.4). We now consider a variant of the program in Listing A.1.[1]

---

[1]Experimentation with the program of Listing A.1 is open-ended and, hopefully, informative with respect to chemotaxis. However, each execu-tion produces extensive output, particularly graphical, which could even be extended for additional insight. Therefore, only one variant is consid-ered next, with the suggestion to the reader that experimentation can be informative, particularly in evaluating and explaining the output.

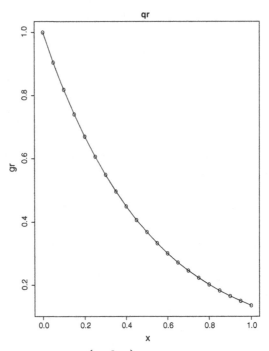

Figure A.1b: $q_r = -\chi_2 \left(m \dfrac{\partial c_2}{\partial x} \right)$ against x line - num, points - exact

(A.2) Analysis of the PDE Chemotaxis Terms

To conclude this discussion of the chemotaxis terms in eqs. (2.3), (2.4), we consider the case of increased effect of the attractant (with concentration $c_1(x,t)$) and repellent (with concentration $c_2(x,t)$. This is accomplished by changing the exponentials in $c_1(x,t)$ or $c_2(x,t)$ to give larger gradients (slopes).

First, b, is changed from -1 to -2.

```
#
# Define dependent variables
  a=-1;b=-2;c=-1;
  chi1=1;chi2=1;
```

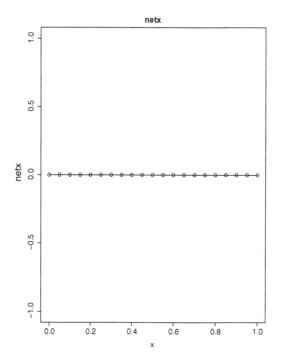

Figure A.1c: $q_a + q_r$ against x line - num, points - exact

```
m=exp(a*x);
c1=exp(b*x);
c2=exp(c*x);
```

The resulting numerical and selected graphical output follows.

x	qa	qr	netx
x	qae	qre	netxe
0.000	-2.00000	1.00000	-0.9999998
0.000	-2.00000	1.00000	-1.0000000
0.250	-0.94473	0.60653	-0.3382025
0.250	-0.94473	0.60653	-0.3382024

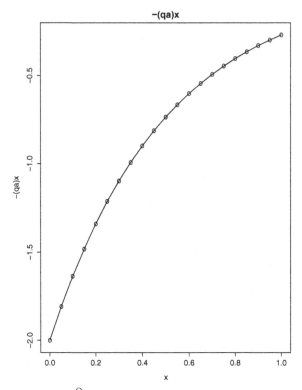

Figure A.1d: $-\dfrac{\partial q_a}{\partial x}$ against x line - num, points - exact

```
0.500   -0.44626    0.36788   -0.0783809
0.500   -0.44626    0.36788   -0.0783809

0.750   -0.21080    0.22313    0.0123317
0.750   -0.21080    0.22313    0.0123317

1.000   -0.09957    0.13534    0.0357612
1.000   -0.09957    0.13534    0.0357611

   x        qax         qrx       netxx
   x        qaxe        qrxe      netxxe
```

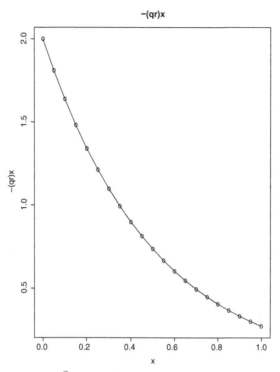

Figure A.1e: $-\dfrac{\partial q_r}{\partial x}$ against x line - num, points - exact

```
0.000    -5.99998    2.00000    -3.9999761
0.000    -6.00000    2.00000    -4.0000000

0.250    -2.83420    1.21306    -1.6211382
0.250    -2.83420    1.21306    -1.6211380

0.500    -1.33878    0.73576    -0.6030222
0.500    -1.33878    0.73576    -0.6030221

0.750    -0.63240    0.44626    -0.1861351
0.750    -0.63240    0.44626    -0.1861350
```

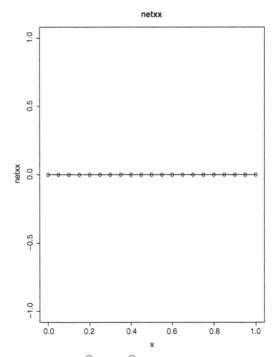

Figure A.1f: Sum of $-\dfrac{\partial q_a}{\partial x} - \dfrac{\partial q_r}{\partial x}$ against x line - num, points - exact

```
1.000   -0.29872   0.27067   -0.0280527
1.000   -0.29872   0.27067   -0.0280518
```

Table A.2: Output from the program of Listing A.1, $b = -2$

We can note the following details about this numerical output.

- The numerical and exact chemotaxis terms again agree to five figures.
- The net flux $q_a + q_r$ is largely negative which means that the net movement of cells with concentration $m(x, t)$ is right to left (in x) into the region of higher $m(x, t)$. Thus, the cell concentration increases (from (eqs. (2.3a),

(2.4a)) with a corresponding increase in $c_1(x,t), c_2(x,t)$ (from
`+a1*m[i]-b1*c1[i]]`, `+a2*m[i]-b2*c2[i]` in Listing 3.1).
Then, an increase in the attractant (increased $c_1(x,t)$) could
lead to aggregation and fibril formation.

```
0.250   -0.94473    0.60653   -0.3382025
0.250   -0.94473    0.60653   -0.3382024
```

- Near $x = 1$, the gradient (slope) of $c_1(x,t)$ is less than that of $c_2(x,t)$ and the net flux changes sign.

```
1.000   -0.09957    0.13534    0.0357612
1.000   -0.09957    0.13534    0.0357611
```

These conclusions are reflected in Figs. A.2a,b,c (note in particular the net flux in Fig. A.2c).

Alternatively, if $c = -2$ thereby giving a larger gradient in
$c_2(x,t)$, the numerical output is

x	qa	qr	netx
x	qae	qre	netxe
0.000	-1.00000	2.00000	0.9999998
0.000	-1.00000	2.00000	1.0000000
0.250	-0.60653	0.94473	0.3382025
0.250	-0.60653	0.94473	0.3382024
0.500	-0.36788	0.44626	0.0783809
0.500	-0.36788	0.44626	0.0783809
0.750	-0.22313	0.21080	-0.0123317
0.750	-0.22313	0.21080	-0.0123317
1.000	-0.13534	0.09957	-0.0357612
1.000	-0.13534	0.09957	-0.0357611

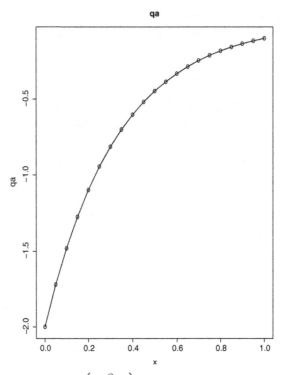

Figure A.2a: $q_a = \chi_1 \left(m \dfrac{\partial c_1}{\partial x} \right)$ against x $b = -2$; line - num, points - exact

x	qax	qrx	netxx
x	qaxe	qrxe	netxxe
0.000	-2.00000	5.99998	3.9999761
0.000	-2.00000	6.00000	4.0000000
0.250	-1.21306	2.83420	1.6211382
0.250	-1.21306	2.83420	1.6211380
0.500	-0.73576	1.33878	0.6030222
0.500	-0.73576	1.33878	0.6030221

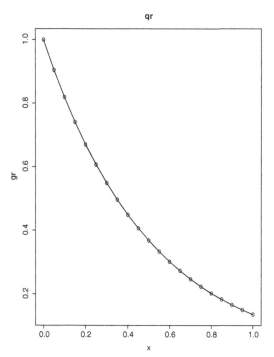

Figure A.2b: $q_r = -\chi_2 \left(m \dfrac{\partial c_2}{\partial x} \right)$ against x $b = -2$; line - num, points - exact

```
0.750   -0.44626    0.63240    0.1861351
0.750   -0.44626    0.63240    0.1861350

1.000   -0.27067    0.29872    0.0280527
1.000   -0.27067    0.29872    0.0280518
```

Table A.3: Output from the program of Listing A.1, $c = -2$

We can note the following details about this numerical output.

- The numerical and exact chemotaxis terms again agree to five figures.

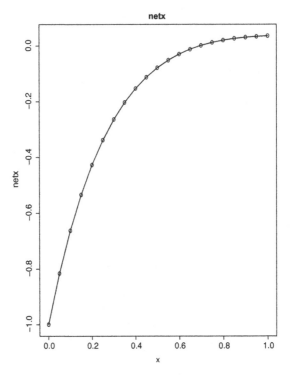

Figure A.2c: $q_a + q_r$ against x $b = -2$; line - num, points - exact

- The net flux $q_a + q_r$ is largely positive which means that the net movement of cells with concentration $m(x, t)$ is left to right (in x) into the region of lower $m(x, t)$. Thus, the cell concentration decreases (from (eqs. (2.3a), (2.4a)) with a corresponding decrease in $c_1(x, t), c_2(x, t)$ (from `+a1*m[i]-b1*c1[i]`, `+a2*m[i]-b2*c2[i]` in Listing 3.1). Then, a decrease in the attractant (decreased $c_1(x, t)$) could lead to lower aggregation and fibril formation.

```
0.250   -0.60653   0.94473   0.3382025
0.250   -0.60653   0.94473   0.3382024
```

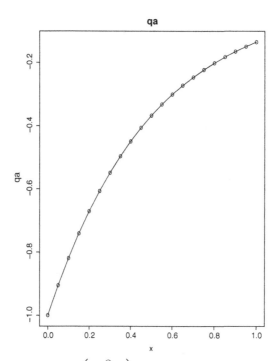

Figure A.3a: $q_a = \chi_1 \left(m \dfrac{\partial c_1}{\partial x} \right)$ against x $c = -2$; line - num, points - exact

- Near $x = 1$, the gradient (slope) of $c_2(x, t)$ is less than that of $c_1(x, t)$ and the net flux changes sign.

```
1.000  -0.13534   0.09957  -0.0357612
1.000  -0.13534   0.09957  -0.0357611
```

These conclusions are reflected in Figs. A.3a,b,c (note in particular the net flux in Fig. A.3c).

These results (for $b = -2$, then $c = -2$) suggest that lowering the attractant concentration is possible through movement according to the chemotaxis equations (2.3), (2.4). This in turn could lower fibril (plaque) formation and thereby reduce the

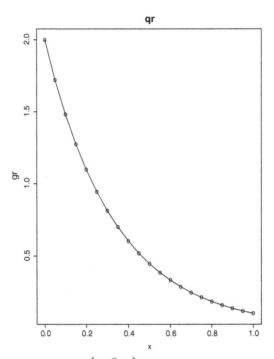

Figure A.3b: $q_r = -\chi_2 \left(m \dfrac{\partial c_2}{\partial x} \right)$ against x $c = -2$; line - num, points - exact

incidence of ND such as AD. This also suggests as a therapy the introduction of a repulsive component to lower the concentration of cells and chemicals (lowering of $m(x,t), c_1(x,t),$ $c_2(x,t)$).

The principal intent of the preceding development and discussion of the chemotaxis model, eqs. (2.3) to (2.6), is to demonstrate how a mathematical model might be useful in understanding a difficult physiological condition. The model and associated R routines can easily be modified and/or reformulated to reflect the interests of the reader. That is, a general methodology has been presented for PDE model development as explained through a specific example.

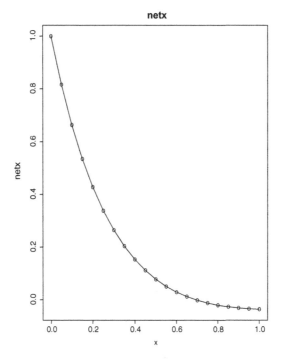

Figure A.3c: $q_a + q_r$ against x $c = -2$; line - num, points - exact

Appendix B: Function dss006

dss006 (Derivative in Space Subroutine) is listed next. The sixth order finite difference (FD) approximations programmed in dss006 were produced with subroutine **weights** by Fornberg [1].

```
  dss006=function(xl,xu,n,u) {
#
# Function dss006 is based on sixth
# order finite differences (FDs) for
# the calculation of a first derivative.
#
# Input argument list
#
#   xl lower boundary value of the
#      independent variable x
#
#   xu upper boundary value of the
#      independent variable x
#
#   n  number of FD grid points
#
#   u  n-vector of dependent variable values
#
# Output
#
#   ux n-vector of the derivatives
```

```
#       of u with respect to x
#
# Preallocate arrays
  ux=rep(0,n);
#
# Grid spacing
  dx=(xu-xl)/(n-1);
#
# 1/(6!*dx) for subsequent use
  r6fdx=1./(720*dx);
#
# ux vector
#
# Boundaries (x=xl,x=xu)
  ux[1]=r6fdx*
    (-1764*u[  1]+4320*u[  2]-5400*u[  3]+
      4800*u[  4]-2700*u[  5] +864*u[  6]-
       120*u[  7]);
  ux[n]=r6fdx*
    ( 1764*u[  n]-4320*u[n-1]+5400*u[n-2]-
      4800*u[n-3]+2700*u[n-4] -864*u[n-5]+
       120*u[n-6]);
#
# dx in from boundaries (x=xl+dx,x=xu-dx)
  ux[  2]=r6fdx*
    (-120*u[  1]-924*u[  2]+1800*u[  3]-1200*u[  4]+
      600*u[  5]-180*u[  6]  +24*u[  7]);
  ux[n-1]=r6fdx*
    ( 120*u[  n]+924*u[n-1]-1800*u[n-2]+1200*u[n-3]-
      600*u[n-4]+180*u[n-5]  -24*u[n-6]);
#
# 2*dx in from boundaries (x=xl+2*dx,x=xu-2*dx)
  ux[  3]=r6fdx*
    ( 24*u[  1]-288*u[  2]-420*u[  3]+960*u[  4]-
```

```
   360*u[  5] +96*u[  6] -12*u[  7]);
 ux[n-2]=r6fdx*
   (-24*u[  n]+288*u[n-1]+420*u[n-2]-960*u[n-3]+
   360*u[n-4] -96*u[n-5] +12*u[n-6]);
#
# Interior points (x=xl+3*dx,...,x=xu-3*dx)
 for(i in 4:(n-3))
 ux[i]=r6fdx*
   (-12*u[i-3]+108*u[i-2]-540*u[i-1]+
   12*u[i+3]-108*u[i+2]+540*u[i+1]);
#
# All points concluded (x=xl,...,x=xu)
 return(c(ux));
}
```

Listing B.1: Function dss006 for numerical first derivatives

We can note the following details about dss006:

- The function is defined.

```
 dss006=function(xl,xu,n,u) {
#
# Function dss006 is based on sixth
# order finite differences (FDs) for
# the calculation of a first derivative.
```

The input arguments are explained in the comments, and their use is illustrated in the previous examples (in Chapters 3, Appendix A). The computed first derivative n-vector is placed in array ux (ux=rep(0,n)).

- A multiplying factor, $\frac{1}{6!\Delta x}$, is computed for use in the FD formulas.

```
#
# Grid spacing
```

```
dx=(xu-xl)/(n-1);
#
# 1/(6!*dx) for subsequent use
r6fdx=1/(720*dx);
```

Note in particular the FD grid spacing $\Delta x =$ dx.

- The FD approximation of $\dfrac{\partial u}{\partial x}$ is computed over the grid of n points with subscripting in i as listed next (with CFD $\Rightarrow$ centered FD, NFD $\Rightarrow$ noncentered FD).

Derivative	Index i	ux[i]
$\dfrac{\partial u(x = x_l, t)}{\partial x}$	1	ux[1] (NFD)
$\dfrac{\partial u(x = x_l + \Delta x, t)}{\partial x}$	2	ux[2] (NFD)
$\dfrac{\partial u(x = x_l + 2\Delta x, t)}{\partial x}$	3	ux[3] (NFD)
$\dfrac{\partial u(x = x_l + (i-1)\Delta x, t)}{\partial x}$	4 to n-3	ux[i] (CFD)
$\dfrac{\partial u(x = x_u - (n-2)\Delta x, t)}{\partial x}$	(n-2)	ux[n-2] (NFD)
$\dfrac{\partial u(x = x_u - (n-1)\Delta x, t)}{\partial x}$	(n-1)	ux[n-1] (NFD)
$\dfrac{\partial u(x = x_u, t)}{\partial x}$	n	ux[n] (NFD)

- NFD approximations are used in the neighborhood of the boundaries $x = x_l, x_u$ to avoid fictitious values beyond the boundaries. For example, ux[1] is a weighted sum of the values u[1],u[2],u[3],u[4],u[5],u[6],u[7] (point 1 and six (interior) points to the right).

- For the interior points ($4 \leq i \leq n - 3$), a CFD is used with weighting coefficients -12, 108, -540, 0, 540, -108, 12. Note that these coefficients are antisymmetric with respect to the center point i (opposite in sign) as required by an odd order derivative, and they sum to zero which is required to correctly differentiate a constant.
- In each of the FD approximations, seven values of the dependent variable u are used in a weighted sum. The resulting numerical derivative is sixth order correct, that is the truncation error is $O(\Delta x^6)$.

The accuracy of the FD approximations is reflected in the numerical derivatives computed in Appendix A. dss008 has a similar structure and the same calling sequence (argument list). It is based on FDs that are a weighted sum of nine values of the dependent variable and is $O(\Delta x^8)$. Thus, the accuracy (order) of the FD approximations can be easily studied, e.g., dss008 can be called in place of dss006.

Reference

[1] Fornberg, B. (1998), Calculation of Weights in Finite Difference Formulas, *SIAM Review*, **40** no. 3, 685–691

Index

Printed in the United States
By Bookmasters